Graphic
Design
Essentials

BLOOMSBURY VISUAL ARTS
Bloomsbury Publishing Plc
50 Bedford Square, London, WC1B 3DP, UK
1385 Broadway, New York, NY 10018, USA

BLOOMSBURY, BLOOMSBURY VISUAL ARTS
and the Diana logo are trademarks of Bloomsbury
Publishing Plc

First edition published in Great Britain in 2009,
by Laurence King Publishing Ltd

This edition published in Great Britain in 2020,
by Bloomsbury Visual Arts
Copyright © Bloomsbury 2020

Joyce Walsh has asserted her right under the
Copyright, Designs and Patents Act, 1988, to be
identified as Author of this work.

For legal purposes the Acknowledgments on p.224
constitute an extension of this copyright page.

Cover design: Lou Dugdale

A catalogue record for this book is available from
the British Library.

Library of Congress Cataloging-in-Publication Data

Names: Walsh, Joyce, author.

Title: Graphic design essentials: with Adobe software /
Joyce Walsh.

Description: Second edition. | London, UK; New York,
NY, USA: Bloomsbury Publishing, 2020. |
Includes bibliographical references and index.

Identifiers: LCCN 2020011466 (print) |
LCCN 2020011467 (ebook) | ISBN 9781350075047
(paperback) | ISBN 9781350075061 (epub) |
ISBN 9781350075078 (pdf)

Subjects: LCSH: Adobe Photoshop. | Adobe Illustrator
(Computer file) | Adobe InDesign (Electronic resource)
| Computer graphics. | Graphic arts—Computer-aided
design.

Classification: LCC T385 .W3654 2020 (print) |
LCC T385 (ebook) | DDC 006.6/86—dc23

LC record available at https://lccn.loc.gov/2020011466
LC ebook record available at https://lccn.loc.
gov/2020011467

ISBN: PB: 978-1-350075-04-7
            ePDF: 978-1-350075-07-8
            eBook: 978-1-350075-06-1

Typeset by Struktur Design
Printed and bound in India

To find out more about our authors and books, visit
www.bloomsbury.com and sign up for our newsletters.

Graphic Design Essentials

With Adobe Software
Second Edition

Joyce Walsh

BLOOMSBURY VISUAL ARTS
LONDON • NEW YORK • OXFORD • NEW DELHI • SYDNEY

Preface

Why Write This Book?

This book is inspired by my own experiences learning about graphic design. I remember being surprised that it's not easy. This is partly because novices must build two different areas of expertise to begin creating good work. In addition to learning the fundamentals of graphic design, we need to learn complex software. My book integrates these two, and this approach helps beginners learn the essentials more quickly, and in a way that reflects the nature of the field. Our creative process involves an integration of design strategy with creative software capabilities.

The goal of this book is to efficiently develop a strong foundation in graphic design and production capabilities, whether readers want to pursue design as a career, or gain skills that will complement other professions. This knowledge applies to a great variety of fields. Recently, in one semester, I had students studying advertising, business, computer science, engineering, a pastry chef, and a man who worked in the mayor's office.

This book can also be used by individuals who want to independently learn design strategies and Adobe software.

What You'll Learn

Throughout, my underlying goal is to help you develop solid design fundamentals coupled with professional software skills including Adobe Illustrator, Photoshop, and InDesign. You'll learn about fonts, colors, images, logos, and layouts to engage and communicate effectively in all media.

How To Use This Book

White pages: Each chapter begins with white pages that show examples of effective graphic designs from all over the world. The text explains why their design strategies are successful and how you can apply them in your own work.

Gray boxes: These boxes within the white pages provide you with design analysis and project development practice to reinforce what you're reading.

Yellow boxes: These provide concise yet professional strategies.

Blue pages: These pages introduce Adobe Photoshop, Illustrator, and InDesign with step-by-step software demonstrations that also reinforce design principles and provide professional tips. The software skills are presented incrementally, beginning with the use of essential tools and becoming progressively more complex. You are encouraged to go through the book in order, as this will comfortably build a useful set of skills.

Those readers who are more advanced are encouraged to read through the blue pages, as they provide design tips and strategies. By the end of the book, you will use Photoshop, Illustrator, and InDesign as professionals do, understanding which software is best for which task. The images you'll use in the software skills can be downloaded from http://www.bloomsbury.com/graphic-design-essentials-9781350075047

No experience is necessary! Let's go ...

About the Author

Joyce Walsh is an educator, writer, and designer. As a professor at Boston University, she has taught over 3,000 students and looks forward to many more. With a passion for design, along with academic and professional backgrounds in both the arts and technology, she develops innovative courses that integrate principles of design with creative software skills. Experienced in all areas of graphic design, her work is featured in international exhibitions and publications.

Joyce wrote the first book to combine design principles with software. This, the second edition, contains the latest in graphics software techniques and hundreds of new globally sourced designs.

Introduction

Lucien Bernhard, Priester matches poster, 1905

This Priester matches poster was considered a work of genius and made Lucien Bernhard famous. While a struggling young artist in Berlin, he won a poster contest with this radically concise design. His poster successfully reduced the style of commercial communication from complex lithographs (Art Nouveau posters) to one word and two matches. Bernhard's bold colors make the visually simple message powerful. He repeated his formula—flat background color, product name, and simple image—for over twenty years. His designs moved graphic communications forward and continue to inspire new designers to communicate with simplified form and bold colors.

Throughout this book you'll notice several historic examples that have helped shape graphic design over the past 120 years. These "hero" designers expressed their passion for type, image, color, and layout in their works. I hope they inspire you as you develop your skills in design.

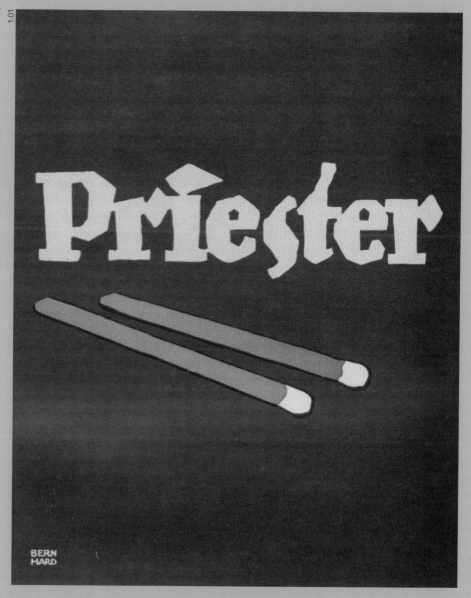

Curves
ahead

An essential tool for curling your lashes
for more definition. Give them body, give them
volume. Because they ask so little in return.

Graphic Design is Everywhere

Look around and you'll see that graphic design is everywhere. It's the most pervasive art form in our society. And when it's good, it's powerful. Graphic design influences our purchases with distinctive branding, clever packaging, and persuasive advertising. It also engages us and enhances our comprehension of text in websites, apps, magazines, and books.

Would you like to produce powerful work such as that shown on this page? This book will teach you how to analyze designs to understand their underlying strategies. It will provide guidelines for successfully choosing and using colors, typefaces, images, and layouts. And it will teach you software skills with design exercises, so you can create your own influential graphic designs in all forms, both print and digital, from package designs to smart watch displays.

Design Concepts
+ Examples
+ Analysis
+ Software Skills
+ Projects

We'll use the formula above throughout the book. Each chapter provides explanations of design concepts, along with examples and analysis that reinforce this knowledge. This information is complemented with instruction in Adobe Creative Cloud software. The software skills are demonstrated with engaging exercises that will reinforce your understanding of design fundamentals. Projects provide you with opportunities for independent creative development using professional design software.

Adobe Creative Cloud
The Adobe Creative Cloud includes industry-standard software: Photoshop, Illustrator, and InDesign. Photoshop is the leading professional software used to optimize photographic and complex images. Illustrator is used to draw on the computer and for single-page layout for print and screen. InDesign is multiple-page layout software

and is used to organize designs prior to sending jobs to print shops or developers for digital products. These three Adobe products share a similar interface that facilitates the beginner's ability to use the software.

A reassuring consideration: in this book software skills are built incrementally. Going through the chapters sequentially, you are provided with software instruction to perform each exercise. The software sections will advance your skills progressively. And the exercises reinforce design concepts, so you'll build design capability along with technical facility.

As you develop these skills, your confidence will grow. You will develop the ability to produce the ideas you have in your imagination. If you already know a bit of Photoshop or other programs, doing the exercises anyway will reinforce the design concepts and build your software skills to a professional level. In the later chapters, you will, as most designers do, combine your skills in all three programs to produce designs. Ready? Let's go!

Functional Fine Art

Designs of words and images are everywhere, and when done well, they're considered functional fine art. The goal of most graphic design is to communicate, but visual appeal can be subjective: what you like may differ from what another person likes. Regardless of style, good design enhances our lives, while bad design impedes communication and comprehension.

1.04

1.05

Think about the graphic designs you've seen today. The examples on these pages represent the range we encounter daily: packaging, signs, logos, magazines, websites, and apps. Logos appear everywhere, from coffee cups to the tops of buildings, and act as stamps of quality and cost. Signs direct us to new locations, saving time or causing more steps. Advertisements, in print or flashing at you on the internet, can affect our purchases, whether we think we want a product or not. Magazines are designed to inform and influence readers.

Graphic designs facilitate our ability to get the information we need, yet we've all experienced poorly organized websites that are incomprehensible. From smartphone icons to book covers, graphic designs influence our behavior for better or worse. The design of this book was carefully constructed to enhance your reading and learning experiences. As we begin Chapter 1: Introduction, let's look at examples and begin analyzing design to discover the underlying strategies for success.

Analyze Design

When the reality program *Survivor* began in the United States, millions were intrigued by this sociological competition and felt compelled to watch sixteen people trying to survive in the wilderness. But it's the logo that captivated me. It conveys a lot of information quickly and effectively. At a glance you see the name of the program and a tropical island environment. Now look closely at the images. They are actually very simple shapes—the top half of the oval has a collection of overlapping palm trees; look even closer, the trees are all the same shape, simply placed at varying angles and sizes to suggest density. Notice how the tips of the palms overlap the oval border—this technique creates depth, increasing visual interest.

Now look at the lower half of the oval: the image that we immediately perceive as island surf is actually one wavy black line against a blue background. It's remarkable what one line can convey. Limiting the number of colors in a logo is advantageous. Here there are only three: black, vivid blue for the water, and green for foliage. Although limited, this color palette immediately provides information about the environment. Next, look at the type. It appears roughhewn or hand-carved, suggesting the castaways' experience. Finally, notice how your eyes follow the entire design in a logical path. You perceived the image and read the text due to effective use of visual hierarchy: the name "Survivor" is largest, then the slogan "outwit, outplay, outlast" is in a simpler, smaller typeface. Your eye is led logically throughout all the information.

The following season, *Survivor* moved to another remote and hazardous location. The logo design strategy remained the same, yet modifications effectively convey the new locale in Australia. The new color palette is warm—orange, yellow and brown. It now looks hot and dry. The illustrations remain simplified—the wavy black line is smoothed, turned 90°, and repeated to suggest the wind-blown dry, red earth in the outback. There's a new graphic element in this logo—notice the notches on either side of the oval. While abstract, these subtly suggest Australian themes, alluding to Aboriginal art or boomerangs. In the logos chapter, we'll learn why simplicity is important in logo design: it enhances recognition and facilitates reproduction.

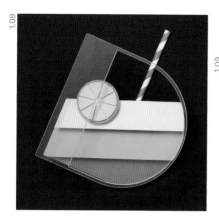

This charming letter D represents the Best of Drinks category in the *Washington Post*'s Readers' Favorites site. The design team built each letter using relevant materials. The clear parts of the D are real glass. The clever use of color creates added appeal. Red and green are complementary colors, and their pale shades work well to suggest a refreshing beverage. A basic principle (contrast) and two elements of design (color and shape) create the witty and effective result. You'll read more on these strategies in Chapter 2: The Elements of Design.

Did this title design catch your eye? At this size, the three letters—ART—demand our attention, as they are so much larger than the rest of the text. This size contrast creates a focal point that pulls us into the story. There are two fonts in this design, with variations of sizes and weights. You'll learn how to effectively choose and use fonts in Chapter 3: Typography.

Illustration is less common than photography in graphic design, but it can be advantageous, as we see in this book cover. Illustration has a timeless quality, while photography can look dated quickly. If your project has a long shelf life, consider alternatives to photography. In Chapter 4: Images, we'll learn when it is advantageous to use illustrations in graphic designs.

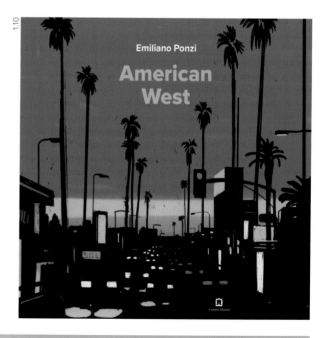

1.11

Designers often use photographs to immediately communicate a message or set a tone. In this spread, we are attracted to the article because of the appetizing photo. So many cheeses! Am I hungry? The designer arranged over twenty foods by grouping them onto four cutting boards. The boards were then arranged in a grid format, so the many details are less overwhelming. In Chapter 4: Images, you'll learn how to choose photographs and reproduce them at their best quality.

1.12

This poster and tote bag use abstract shapes and colors to attract our attention. Abstract images can be beneficial because they may represent many different possibilities, and the audience can choose their own interpretations. Straightforward type, bold colors and a handmade quality suggest a contemporary and creative community at Second Home. The poster is a centered or symmetrical layout. The tote bag has an asymmetrical placement of the text and images. We will explore symmetrical and asymmetrically balanced designs in the Chapter 5: Layouts.

1.13

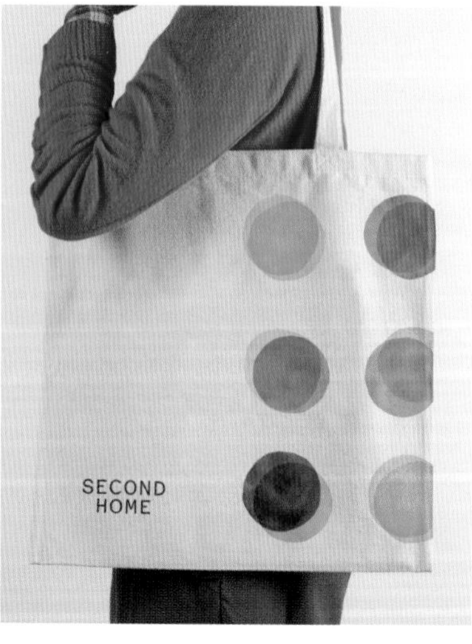

This book attracts our attention with the unusual direction of the title that runs along a suburban road. The illustration includes bird's-eye view of houses. White rectangles that look like driveways are used to showcase the author's name and subtitles. All the text aligns with white backgrounds that help complete the illustration with roads and driveways. In Chapter 5: Layouts, you'll learn the value of integrating type and image to ensure your text will be read.

1.14

Who doesn't love to pick up matchboxes as a memento of a night out? Simon & The Whale is a restaurant in New York City that draws menu inspiration from both coasts of the United States. The colors and handmade style of the logo evoke the relaxed and nostalgic mood you'll find at the restaurant. You'll learn brand design strategies in Chapter 6: Logos.

Many, if not most, graphic designs require multiple formats, as we see in these Shake Shack packages. The Shake Shack logo is recognized around the world. It is a flexible system we'll see more of in Chapter 7: Visual Themes. You'll learn how to create visual themes and apply repetition and variation to achieve cohesion, consistency, and rhythm throughout your designs.

Many graphic designs are multiple-page productions, such as websites, magazines, books, and annual reports. Printed or digital, the design cannot be understood at a glance, and the audience must click or flip through the pages for the complete experience. What are the similarities on the two pages of this Desk Plants brand guideline book? Consistency of fonts, image style, and layout provide unity and enhance comprehension of the publication. What are the slight differences on these two pages? The designer's challenge is to be visually cohesive, without falling into redundancy, throughout a publication or website. Strategies for achieving successful visual themes are covered in the last chapter. All the information and skills you've learned will come together in

Chapter 7: Visual Themes.

Graphic design influences our impressions everywhere we go. Both locally and when we travel abroad, most places have a visual character established by factors like their climate and architecture. Designers use these traits when creating environmental graphics, such as signs, wayfinding systems, and banners. Environmental graphics are physical, rather than digital, and have an impact on real life.

The Cochon Dingue (crazy pig) sign contrasts comfortably with the traditional architectural details of the building. The wrought-iron sign hanger marries the contemporary design to the traditional architecture. The awning and shutters are the same blue as in the logo, inspired by the French flag. You'll read this beloved restaurant's brand update in the logo chapter.

This sign is part of the wayfinding system for a riverside precinct in Queensland, Australia. Wayfinding designs consider humans' intuitive method for finding places, and supplement that with signs, pathways, lighting, and gathering places. This site's strategy encourages circulation and draws people down to the banks of the river. The identification and directional signs also provide history about the area. This area was rebuilt after a cyclone in 2015. Can you see how the materials were selected and designed for weather resistance?

Waltzing Matilda is a museum in the Australian outback. The rusted steel that forms the structure of the sign draws from the stark, arid landscape. The rugged, straightforward typeface reflects the area's stalwart identity. These materials were inspired by the outback's rock formations, prehistoric craters and warm colors.

We commonly see banners hanging in our communities; these are semi-permanent environmental graphics. Ideally, they reflect their locales. Real estate company Halstead wisely chose blue for this banner color. It is the Western world's most common favorite color, and blue conveys security and orderliness. The sans-serif font and angular logo also set a professional tone. Real estate transactions are often the largest financial investment people make; these details help to evoke confidence in the firm. What color might you avoid using for Halstead projects?

Design Analysis: Environmental Graphics

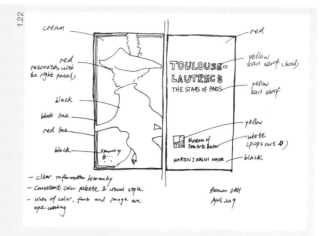

Environmental graphics are relatively permanent graphic designs that influence our impressions of a location. You'll see them on lampposts, buildings, and stand-alone signage, sometimes called street furniture. Take a walk and find two local examples of environmental graphics, one good and one less successful. Consider what details lead to successful designs or unsuccessful designs. Make a sketch of each using a black marker.

Objective

This exercise has multiple objectives. It reinforces the idea that graphic design is the most common form of art in our society, it provides an opportunity for analysis, and it allows you to become comfortable with sketching, so you can better communicate visual ideas.

To Do

Make sketches of two examples of environmental graphics, one good and one bad.

The primary goal is to start looking, analyzing and sketching. When we have this conversation in class, there's always a banner (including these) that are considered good by some and bad by others. As you go about your day, continue to analyze the many types of graphic design you encounter, noting how each design communicates its message. As you progress through this book, you'll learn about the underlying design strategies of successful projects. In the next section of this chapter, we'll begin to develop your software skills, another essential component in creating effective graphic designs.

Sketches by students
Yuling Lu and Val Su

Tip

Sketch each banner while observing—don't trust your visual memory. The sketch should show the outside proportion of the banners, the image, type style, and layout. Make the sketches with a black marker and add notes to indicate colors. This conveys the design without need for a careful rendering. The goal is to share visual ideas without having to make a time-consuming illustration. As many of us aren't skilled illustrators, this approach takes the pressure off.

Introduction to Adobe Photoshop

Photoshop is used to modify photographs and elaborate illustrations for use in designs. For every task you do in Photoshop, there are at least two ways to accomplish it, but my approach is to provide you with a method that leads to professional-quality results. If you've been using computers since early childhood, you'll understand many aspects of new software intuitively. In these exercises I will teach you the less intuitive Photoshop techniques.

The skills that you'll develop in these exercises will enable you to optimize the appearance of photographs and allow you to make realistic, surrealistic, and sometimes playful modifications to photographs. Photoshop's

capabilities are vast—even full-time professionals don't have the opportunity to explore every niche—so next we'll concentrate on the tools that will be most useful as you begin your design career.

Photoshop Tools

Let's get to know the software and start on the left with the toolbar. The most essential tools for achieving professional results are labelled. In the top position, the Move tool is used to reposition objects on the page.

In Photoshop, areas of an image may need to be selected before you can change them. There are three primary tools for selecting. The Marquee tool selects geometric areas, usually rectangular or circular. Just below is the Lasso tool, which can select organic shapes. Next is the Quick Selection tool; however,

when you hold the mouse button down, you can select the more professional Magic Wand tool, which detects similar colors and groups them into a useful selection.

The crop tool is used to permanently change the image format. The very helpful Eyedropper selects colors from a photo to be used elsewhere, such as in text. The Clone tool looks like a rubber stamp and has impressive copying capabilities, use this tool to make professional-quality image edits. The Paint Bucket hides behind the Gradient tool and is handy for filling areas with a color. The Dodge and Burn tools lighten or darken the exposure of specific areas of a photo. The Type tool functions similarly to typical type tools; differences arise in the display of type in rasterized images.

When chosen, all the tools have

custom settings that are displayed in the Options bar (at the top of the window) or the Properties panel (on the right). These options change according to the selected tool. Keep an eye on these options or properties as you use the various tools.

At the bottom of the toolbar, the color squares functions are unique to Photoshop. On the top is the Foreground Color, the lower square is the Background Color. As with all Adobe software we'll use in this book, the tiny button of black-and-white squares allows you to quickly change to default colors; the bent arrow is used to switch the foreground and background colors.

The Zoom tool behaves just like you'd expect; if not, turn off the Scrubby zoom box that appears in the Options bar when the Zoom tool is selected. You can also use keyboard shortcuts for zooming.

Layers: A Brilliant Feature

Photoshop layers are wonderful—multiple layers allow you to make changes to specific areas of your image without affecting other areas. A new layer is created every time you paste an object or add text to a file. This allows you to edit images and format text or apply transparency (or other effects) to certain layers only.

Make changes to a layer by clicking on the layer shown in the window. The layer is active when it appears highlighted or gray. Double-click on a layer in the window to apply layer styles such as shadows.

View or hide specific layers by clicking the eye icon in the left-hand column of the window.

Once you have completed a design, you can choose to merge layers to reduce file size using the Layers panel menu. You'll learn to use layers in the first Photoshop Software Skills pages.

We work in the default Essentials Workspace. Set it under Window > Workspace > Essentials.

If at some point you can't find a window or panel, simply reset the workspace by choosing Window > Workspace > Reset Essentials.

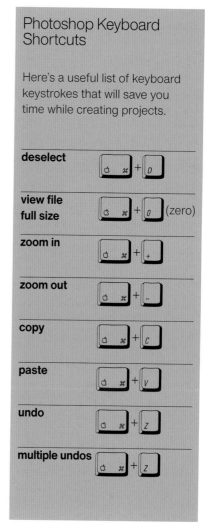

Photoshop Keyboard Shortcuts

Here's a useful list of keyboard keystrokes that will save you time while creating projects.

deselect	⌘ + D
view file full size	⌘ + 0 (zero)
zoom in	⌘ + +
zoom out	⌘ + -
copy	⌘ + C
paste	⌘ + V
undo	⌘ + Z
multiple undos	⌘ + Z

Design + Software Skills 1:
Photoshop Layers, Lasso, Image Adjustments

The software exercises throughout the book will build technical skills incrementally. You will quickly go from basics, such as keyboard shortcuts, to complex layouts, so do the exercises in the order they appear. The software instruction also reinforces design fundamentals and provides many tips for creating professional-quality work.

Objectives

This exercise has several objectives. You will build Photoshop skills for tools that are not intuitively understood, such as the Lasso and Polygonal Lasso; learn how to use layers; resize images proportionately; and useful keyboard shortcuts.

To Do

Open two images, copy one into the other. Learn to use the essential Lasso tool to select a figure, resize it, and put it in the driver's seat.

1

Download the photos

Download two images to your desktop, Convertible.jpg and Archie.jpg from http://www.bloomsbury.com/graphic-design-essentials-9781350075047

2

Open the files

Start Photoshop. Open the Convertible image file by choosing File > Open, then click on Convertible.jpg and press Open. Open the Archie photo the same way.

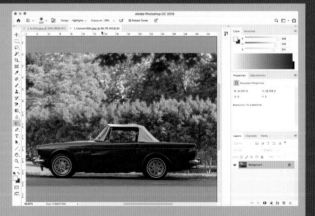

3 Copy

Look at the top of the window, click on the Archie.jpg tab. Press Command + A to select the entire image. The dotted line around the image indicates it is selected.

Now, click on the Convertible.jpg tab. Press Command + V to copy the Archie file into the convertible file.

Find your Layers window in the lower right. Click on the layer with the dog thumbnail image.

4 Layers

Notice there are two layers—this is a good thing! Layers allow you to edit separately.

Click on the Eye icon on the lower, background layer so it is invisible. Click on the dog layer so you can edit it.

5 Lasso

Select the dog with the Lasso tool. First, confirm that you have the regular Lasso tool selected—this is the best choice for selecting outlines of organic shapes. The other Lasso options are the Polygonal Lasso tool, which is used to select straight edges, and the Magnetic Lasso tool, which is used to drive you insane. The magnetic version is one of those software tools that promises more than it delivers. In most instances, the regular Lasso and Polygonal Lasso will work more reliably than the Magnetic Lasso.

With the Lasso tool, click at the outer edge of Archie's left ear; keep holding your mouse button down as you drag around the dog's figure. No need for perfection here, it takes practice. Trace the entire figure, finally ending at the beginning point. Once completely around the dog, take your finger off the mouse. You should see that the entire figure is surrounded by a dotted line. This indicates that the object is selected.

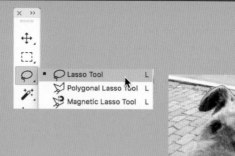

6 Copy and paste

While the dog is selected, press Command + C, then Command + V. This makes a copy in a new layer. Toggle off the visibility of Layer 1 by clicking on the Eye icon. Notice there is a gray checkerboard pattern indicating transparency. This is because of the lasso selection.

Be sure to click on Layer 2 to make it active; it will be gray, as you see here.

7 Resize

First, toggle on the visibility of the Background layer by clicking on the Eye icon. This will help us to scale Archie to fit the car.

Select the Move tool at the top of the Toolbar. Confirm you are on Layer 2—it should be gray.

Resize the dog proportionately. Choose Edit > Transform > Scale. Use a corner handle and drag towards the center to make him smaller, then press Return. When resizing, handles appear around the figure. Press Return to see the resizing results and the handles disappear.

Click on the dog and drag him to the car window. Position him just below the window edge of the door. We're going to let his chin hang out. Adjust the scale, if necessary, using a corner handle.

Tip

Last year, Photoshop made proportionate resizing the default; there's no need to hold the Shift key down in Photoshop to resize proportionately anymore! If you do hold the Shift key down, it will allow disproportionate scaling.

8 Zoom in

Use the Zoom tool to get a close look at the dog. Select the Zoom tool, place the cursor near the top left, hold the mouse button down, and drag to the lower right of the dog. Let go of the mouse button. You are now zoomed directly to the appropriate area of the image.

Tip

Anytime you want to see your entire image, press Command + 0.

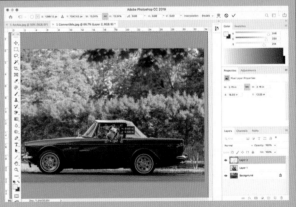

9 Trim

Adjust the Opacity setting temporarily on Layer 2 to about 80 percent. Now we can see the edge of the car window as we fine-tune this masterpiece.

Select the Lasso tool. In the Options bar, change the Feather setting to 1 px.

Select around Archie's beard, allowing his chin to hang out of the car. Press Delete. Press Command + D to deselect.

If necessary, you can zoom in again and use the Lasso tool to trim away the original background from Archie's face.

10 Trim

Now switch to the Polygonal Lasso to select a straight edge at the window. Notice the Feather setting goes to 0 px. This results in a sharp edge. Press Delete and then Command + D to Deselect. Change the Opacity percentage to 100 for Layer 2.

Tip

Typically, I use the Lasso with 0 px Feather edge for sharp-edge results. This image is an exception because of the furry subject.

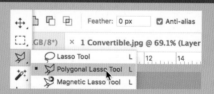

11 View full size
Press Command + 0 to show full size and admire your results.

12 Adjust exposure
Should you find a need to adjust a photograph's exposure, Photoshop provides automatic adjustments under the Image > Adjustments menu. Rather than relying on the auto-corrections, the best tool at this point for adjusting exposure is the Curves feature. Choose Image > Adjustments > Curves.

13 Curves window
Confirm that the Channel selection is RGB—this setting adjusts all colors in the image. Confirm that the lower left-hand corner bars display black.

To make the image slightly darker to match the car lighting, click on the line in the upper-right quadrant. Keep your finger on the mouse and tug ever so slightly down. Notice the change in the image—a very slight movement on the curve darkens the exposure.

To make your image lighter, click on the line in the lower-left quadrant. Keep your finger on the mouse and tug ever so slightly up. Notice the change in the image. A very slight movement on the curve makes significant changes to lighten the exposure.

When satisfied with your adjustments, press OK. To start over from the original image, you can press Cancel. Or, press Command + Z and reselect Image > Adjustments > Curves.

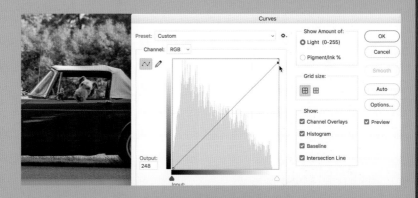

14 Practice

Save this work as a Photoshop file. Select File > Save As and choose Photoshop, the top option on the menu. That will preserve all your layers.

Now put yourself in the driver's seat. Open a head shot of yourself, copy it into this Photoshop file, and go through steps 1–13 again. Great work! Next, try your skills with a different vehicle; here I am in my dream-car.

Major Points Summary

— Graphic design is everywhere.

— Examples include packaging, logos, signs, advertisements, magazines, apps, books, websites ... The list goes on.

— All are forms of communication.

— Graphic design can influence our actions, purchases, and the way we live.

— Good design is functional fine art; it enhances our lives.

— Logos require simplified forms for effective recognition, recall, and reproduction.

— Contrast is used to attract attention to a design.

— Visual hierarchy created with typeface size and style choices improves comprehension.

— Layouts organize content to better communicate the message.

— Integrating type and image ensures the message will be read.

— Visual themes—consistent use of fonts, colors, and style of images—provide cohesion for multiple-page designs, such as apps, websites, ad campaigns, and magazines.

— Sketches help to communicate visual concepts.

— For accuracy, sketch while observing the object.

Software Skills Summary

Photoshop Introduction

Overview of the toolbox and the most commonly used tools for beginning designers: Move, Marquee/Ellipse, Lasso, Quick Selection/Magic Wand, Crop, Clone, Gradient/Paint Bucket, Dodge/Burn, Type, Eyedropper, Zoom, Foreground and Background Colors, Color Toggle, and Default Colors, Image, Adjustments, Curves.

Keyboard shortcuts.

Skills: Layers, proportionate resizing, Options bar, curves adjustments.

Tools covered in depth: Zoom, Lasso, Move, Magic Wand.

Recommended Readings

Each chapter combines design concepts + examples + analysis + software skills + projects on the elements of design, type, images, layout, logos, and visual themes. Additionally, every chapter will provide you with further reading or listening suggestions for each topic. To see more inspiring design, look for design annuals. These are published results of international competitions for the year's best design. Designers use these books, magazines, catalogs, websites, and even podcasts, for inspiration and to keep up with trends and the studios that are creating great work. The magazine *Communication Arts* publishes a highly anticipated graphic design edition every autumn. Its interactive annual is published in the spring, and all annuals are on their website: https://www.commarts.com/.

For advice and stories from successful designers, listen to Debbie Milman's podcast: https://www.designmattersmedia.com/designmatters.

Read Adrian Shaughnessy's classic, *How to Be a Graphic Designer Without Losing Your Soul*.

Another excellent resource is Beauty by Stefan Sagmeister and *Jessica* Walsh. This book explores design, philosophy, history, and science to understand how beauty impacts our lives.

Thank you to Joshua Sweeney, Shoot for Details, for this photo of the Sunbeam Tiger.

A complete list of the contributing designers and citations for books can be found in the Appendix.

The Elements of Design

Michael Schwab, Golden Gate National Park posters, 1995

Michael Schwab's signature style provides timeless images that convey the drama and poetry of each Golden Gate National Park. These posters raised awareness and funding to support the costs of maintaining the parks. Raised on a farm in the Midwest, Schwab studied graphic design in Texas and New York, and he works in San Francisco. His recognizable and often-imitated style is minimalist, with the essence of each subject depicted in flat-black shapes, then enhanced with bold flat color. Notice how color sets a meaningful tone in each of these posters. The red sky over Alcatraz, an infamous prison, signals danger. What do the colors of the skies in the other posters convey?

2.01

CHARLES
BUKOWSKI
Il canto dei folli
Poesie II

UNIVERSALE
ECONOMICA
FELTRINELLI

Creating Powerful Designs

Emilio Ponzi's book cover illustration captures our attention with his clever use of design elements, especially color. Here, he contrasts warm and cool colors. Tans suggest dessert heat (notice the dozing dog). While the red title contrasts with pale greens and blues. Read on to learn the underlying strategies for using the elements of design to create engaging work.

Designers manipulate any or all of the elements of design when developing projects. An element is one of the simplest principles of an area of study. In graphic design, the seven elements are: color; direction; line; size; shape; texture, and value.

Lines

Let's start with a design exercise. Draw a rectangle, any size. Easy for you, right? Now, draw two lines. Are you hesitating? Suddenly, drawing two lines becomes a difficult decision. You've probably not thought much about lines before, but they are a basic building block of graphic design. In other words, an element of design.

Experienced designers know that all designs begin with four lines—the edges of the screen or page that determine the format. Of course, drawn lines are also used in graphic designs. The formal definition illustrates how varied lines can be: a line is simply a mark made by a tool as it is moved across a surface.

Because lines are fundamental, yet so varied, the ability to describe them effectively will lead you to more successful designs. There are a great variety of lines, but they all have three attributes: type, direction, and quality. Type refers to whether it is straight, curved or angular. Direction describes the line's virtual movement on the page. If your lines are horizontal,

you can describe them as running east or west. Vertical lines can be described as going north or south. Indulge your descriptive abilities when describing the line quality. Are your lines thick, bold, or stout; thin, narrow, or slender; wavy, rippling, or undulating? A regular line quality maintains its thickness throughout. The quality of a line drawn with a slanted calligraphy pen nib will change, going from thick to thin along its length. Look at the two lines you drew and describe them using these terms.

Some successful designs use lines, but no color or images, to produce intriguing results. This book cover uses line as abstract form to convey meaning. Columbia University's School of Architecture, Planning, and Preservation published this collection of selected student projects with a restrained cover design expressing many aspects of urban planning. What do these lines suggest to you? Pull the elastic that is the black straight line to reveal the title.

The High Line is a unique park in New York City that was developed

along abandoned, elevated railroad tracks that run along the west edge of Manhattan. It quickly became a favourite destination for New Yorkers and tourists to stroll through gardens planted into the old tracks while enjoying Hudson River views and breezes. Using only three lines, Paula Scher designed this initial H logo suggesting train tracks. Her goal was to avoid an illustrative design and to make it look more like tracks than an H.

Sidekick Solutions provides computer support services, and technicians leave these cards at clients' workspaces when their work is completed. Their logo symbol has two lines: one curved up and one down. These combine to create an S, but do you also see a smile? And on the other curve, do you see a round forehead atop two eyes, perhaps thinking of the solution?

2.04

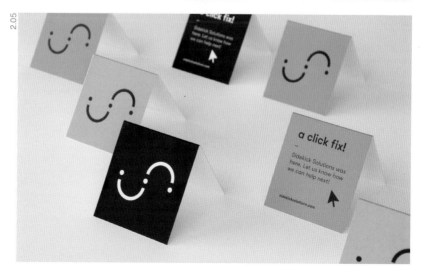

2.05

Shape

Using lines, designers translate the three-dimensional world into shapes. Shape is the general outline of something, whether the object is tangible or abstract. Shapes can also be defined as a closed form or closed path. They can be filled with color, texture, or tone, and they may appear to be flat or have volume.

The clever use of these immediately recognizable shapes draws our attention to this theater poster. *Lasso of Truth* is a play about the origin of Wonder Woman, and the three main characters are playfully represented by these similar, yet distinctive, shapes.

The shapes in this Nike Instagram design are created by the lines on a basketball court. These shapes and the background photos suggest the energy and excitement of the game.

Using minimal color and iconic shapes of dress-forms, hairstyle, croquet mallets, and balls, the play poster hints at the humor in this musical comedy adaptation of the classic mean-girl movie, *Heathers*. There's a cut-throat croquet scene in the movie. Coincidentally, the color pink and similar dress are darkly comic themes from the straight forwardly titled movie *Mean Girls*.

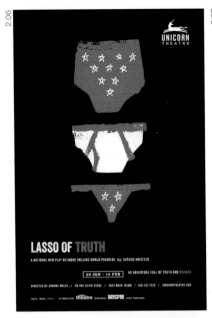

2.06

2.07

2.08

2.09

Value

Value, in visual art, is defined as the relative lightness or darkness of a color. Bloomsbury's *Object Lessons* is a series of books that feature an iconic object. These high heels are created with only three values, 100 percent white, 100 percent black and gray (50 percent black). Additionally, the use of value achieves effective visual hierarchy in the text. We see the 'high heel' white text first because it has the most contrast with the black background. The author's name in blue text is read second because it has less contrast with the black background. The other text, "Object Lessons" and "Bloomsbury", are smaller, thinner letters, providing the least contrast, and are placed on the peripheries to be read last.

The landscape for this book cover is created with several values of blue-green, emphasizing the cold location. The tallest mountain peak has an almost white value that suggests snow. Light values of colors are called tints and dark values are called shades. The author's name is an almost white tint, which creates contrast and

draws attention. Contrast is a term that describes the relative emphasis of an element used in a design.

Because the title is a dark value and smaller, creating less contrast with the background, we see the title secondarily. The use of contrast with the elements' value and size affects how we view the details of this design. The two figures in the illustration wear yellow and orange coats. Even though these are contrasting colors, they are smaller figures, so we may see them after we read the text. Typically, designers create one important area to have the most contrast and to become the primary focal point. Other smaller and less-contrasting areas can attract our eyes in a sequence that moves us through an entire design.

Bring Your Game's text is distorted to make a camouflage pattern in the next poster. Value is the design element that gives this image detail and texture. The values of this camouflage pattern are either 100 percent white or 100 percent black. Teasing through the open areas is a photo of basketball great LeBron James. Many gray values combine to create his profile and figure. Notice the values are stronger in his face and less so in his figure, a contrast drawing

2.11

our eyes to his face. Overall, the foreground has high value contrast, and the photograph has relatively low-value contrasting. These careful design decisions affect how we view this design.

Contrast is often used to achieve greater functionality in graphic design. The words in this book, created with black ink on white paper, have high contrast value to achieve the greatest legibility. Likewise, the digital version of this book, websites, and apps use black text on white backgrounds to achieve highest contrast and best legibility on screens.

2.10

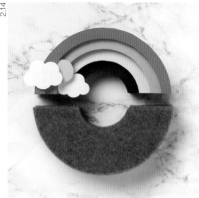

Texture

Texture in graphic design adds visual interest and conveys information. Some visual textures suggest inviting surfaces such as photos of rose petals or skin. Often, these types of images in advertisements suggest creamy products. Visual textures can also appear to be rough, like rust or sandpaper, to express a gritty quality. Here, we see contrasting textures: soft ice cream atop crunchy-looking cones look irresistible. And yes, that's a fish-shaped cone that doesn't taste fishy.

Posters for the Forward Festival provide a frayed, outdoor visual texture, suggesting they have been weathered by the elements.

This effect expresses that the design process always involves construction and deconstruction.

This handmade O represents the category Outdoor Events for the *Washington Post*'s Favorites feature. Notice the grass texture contrasts with the smooth rainbow, and its red and pinks contrast with the greens on the bottom.

The cover image for "Fate" is a large rope knot that conveys a rough texture. This ominous effect is tempered with the pink background and the matte silver varnish, which adds subtle contrasting texture. These visual and tactile textures combine to create a mysterious book cover.

Graphic designers sometimes use tactile effects—real textures—to add unique qualities to designs. Tactile treatments created with embosses, die cuts or folds increase the probability of engaging your audience. Such effects encourage the viewer to pick up and investigate the piece, improving the likelihood that the client's message will be read. Blind embossing, created when a die is pressed into paper, subtly raises the A-shaped logo on the Aslan Foundation business card. The effect implies this organization has a thoughtful mission that perhaps is not obvious at first glance.

2.17

2.20

The Swag Shop sign is made with weathered steel and cutouts. These textures are features of the wayfinding system for this museum in the Australian outback. The materials and design were inspired by the stark, rugged landscapes of the bush.

The title on the book cover for *The Mirror* is reversed. This mirroring effect is enhanced with the use of silver metallic ink. This design uses two elements: texture and direction.

2.19

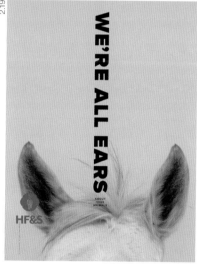

Direction

When an object is placed in an atypical direction, the resulting design engages the audience with a visual puzzle. This is an advertisement for a feed and supply company. The vertical direction of the headline is emphasized by its position between the horse's ears. Other posters in the series include dog, cat, and rabbit ears, as will be seen in Chapter 4 when we discuss photo cropping.

A head-standing skateboarder is the largest image on this two-page title spread, and she is the subject of the article. Her unusual direction captures our attention, and we are engaged with the playfully active collage in which letters and images intersect. Notice how one leg is behind the I's, and the other is in front. This detail helps create depth and visual interest.

2.18

HIGH
LINE
SECTION

OPENING IN JUNE

In the coming weeks, when Section 2 of the High Line opens to the public, history will be made once again. For the first time, visitors will be able to walk the High Line from its southern terminus at Gansevoort Street in the Meatpacking District, through West Chelsea, to West 30th Street at the edge of Hell's Kitchen—an entire mile above the streets of New York City.

The opening of Section 2 means our gardeners will tend over 50,000 new plants and care for a new 4,900-square-foot lawn, and our maintenance team will keep an additional half mile of park clean and safe. As the High Line doubles in size, your continued support will help us operate and maintain the High Line at the highest standards possible.

Thank you to the Friends of the High Line foundation. The Friends for continuing support of the High Line.

Size

This enormous "2" is inescapable for its atypical size and the use of a single digit as a title. For the new designer, using size as an element in design is relatively easy. Simply by enlarging the first letter at the beginning of an article, you have introduced contrast with the body text. This creates a focal point, an entry into the text, and is a typical strategy used in magazines. You will also use size to indicate hierarchy of information—variations show the relative importance of content.

Enormous letters are so big they don't fit on this tote bag, creating dynamic energy with a visual puzzle. Most will immediately recognize the brand, but if not, there is a much smaller black-and-white name towards the bottom left. This creates effective visual hierarchy of information: we initially notice the large blue letters, and secondarily we notice the full name in smaller, regular weight, black-and-white text. You'll read more about this brand re-design in Chapter 6.

Size is often applied to images. Notice a few of the images in this collection of spreads from House Industries are enlarged and closely cropped. When we closely crop images, we achieve two goals: it allows viewers to examine details of the objects, and it provides fresh takes of common objects. Notice that some of these spreads make titles or letters very large, creating visual interest and focal points, while also highlighting the font designs.

Unity

These playful and elegant two-page spreads demonstrate the effectiveness of both unity and contrast in a design. Notice the harmonious use of the same font, the consistent bicycle theme and matte copper ink throughout many of the spreads. These details demonstrate the appeal of unity. Unity is a basic principle of graphic design and dictates that all of the objects in a design be visually harmonious such as on these spreads and throughout this catalog. Now note the contrasts in the spreads: some text is horizontal, some vertical, some photos are smaller and in grids, others are intense close-ups, and large numbers and titles pop-up throughout. These design strategies achieve a visually consistent and compelling design.

Designers must be aware of how they handle formal elements. Whether you're designing a logo or a website, the formal elements are always the same. The elements are interdependent and interact with one another on the page—note when the overall design achieves unity and when contrast captures your attention.

We will explore the element of color in depth in the next section of this chapter.

2.23

Design Analysis: Elements of Design

Objectives

You will identify and analyze the use of the elements of design while reinforcing the concepts introduced in this chapter.

To Do

Find seven examples of designs in which a particular element of graphic design is used to create visual interest. Place these examples in your notebook.

Color

Color Connotations

Color has an immediate effect on its audience. Before type is read and the image is understood, color makes an impression. Red is passionate, whether in love or fury, and carries a strong visual message. It is favoured for designs ranging from national flags to sport cars. This theater booklet's red calendar is noticeable among dark pages, creating an effective design strategy. Because red is so strong, its intensity sometimes needs to be balanced with black to produce a more accessible effect.

Warm colors other than red can attract attention while feeling more innovative. Halstead's brand is designed to stand out in the very competitive New York real estate market. These window banners are inescapable, as their hot-pink text creates a textured pattern on the orange background. The panels are a welcome change from the usual brown paper that covers windows when a space is for rent, and the effect is contemporary and memorable.

Blue has many positive connotations. Dark values of blue, such as navy, suggest expertise, authority, and a seriousness of purpose. Medium values of blue are associated with honesty and cleanliness. Look at the cleaning products on grocery-store shelves to confirm this point.

Yellow is associated with warmth and wholesomeness, so it is often used for food-product designs to convey healthful messages.

Color Combinations

Color is a very powerful and provocative design element, but choosing multiple colors successfully can be challenging. Let's start with an understanding of color theory with a focus on the three primary colors (red, yellow, and blue) and the secondary colors (orange, green, and purple).

When these two groups are placed in overlapping triangles to form a circle, the relationships between the colors become apparent. Colors that appear opposite each other on the wheel are called complementary colors. Opposing positions are significant because these pairs of colors have the most contrast when used together. Contrast creates a visual dynamic that is often desirable in graphic design. The complementary color pairs are red and green, blue and orange, and purple and yellow. The following examples and design exercises will help you to recognize the visual power of these color combinations.

Green is also healthful, indicating freshness and the outdoors. As a term, green represents environmentally sustainable products, practices, and organizations. The Grounds is an oasis in the city of Sydney. Their café, gardens with roaming farm animals, and great coffee provide memorable experiences. An illustration of Fluffy, their resident macaw, graces a distinctive package for their coffee. In addition to the charming illustration, the use of green and blue conveys their commitment to sustainability of the environment in all of the Grounds' enterprises. See more color connotations in the Appendix.

Some people have a natural facility for using color, but for most of us, it can also be learned. Avoid using only your favourites, and select colors that are appropriate for the design. Would you recommend the use of hot pink to a client whose business is financial planning? Probably not: for a client in a conservative industry, it might be better to choose darker colors for immediate and appropriate impact. Pastel colors are commonly used for children's products and services, while deeper shades are considered more conservative. Observe when a design catches your eye and notice which colors are used to create particular effects.

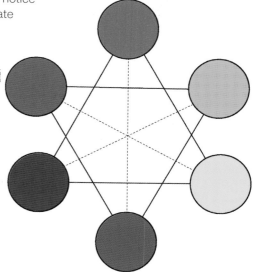

Hot and Cold

Blue and green hues suggest something cool and soothing, while red and yellow suggest warmth, even heat. Red, orange, and yellow are called warm colors; blue, green, and purple are cool colors. Designers often use more of the cool colors and less of the warm colors. Why? The warm colors do attract attention, but for a variety of reasons, too much is not always good thing. On computer screens, too much red can irritate eyes, and on white paper, yellow doesn't have enough contrast to be legible. Blue and green have many positive connotations, such as cleanliness and reliability, and they have good contrast with white backgrounds that are common in all media. It is often wise to use cool hues generously. We will learn to combine cool and warm colors for dynamic designs.

Few color interpretations are as universally agreed upon as warm colors and cool colors. This is likely due to our global human experience that the sun provides warmth, while water and foliage provide cooling effects.

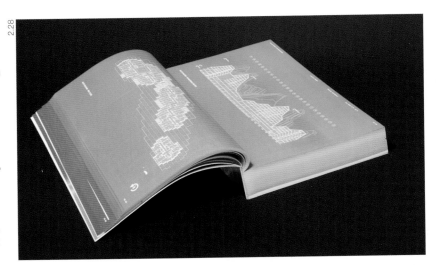

2.28

2.29

This unusual book, *Hot to Cold*, examines sixty architectural sites from the hottest and coldest places on earth. Notice how the colors of the pages change from warm to cool as the topics cover different climatic conditions. The building diagrammed on the orange pages is in a tropical rainforest. The buildings on the blue page are in Sweden. Depending on the ambient temperature, the book cover appears orange and the title says HOT, or the cover is blue with COLD appearing as the title.

Blue and Orange

In this photo of a room at NYU School of Law, cool blue chairs are dramatically set off against short orange walls. The contrast ensures you'll be engaged enough to notice the locations and then read the value of diversity message at the bottom left.

Lateral Cooking used only two ink colors throughout the book: orange is used to highlight page references, footnotes, and sidenotes, and blue is overprinted within the illustrated chapter openers and on the endpapers. This cookbook encourages people to expand their repertoire without the need for more recipes. The design uses reference book conventions, such as the orange step detail, which allows easy navigation to a specific section in a lengthy publication.

This field of business cards pops most dramatically when placed against a blue background. Notice the saturation of both colors is similarly strong. These boldly contrasting hues get our attention.

2.30

2.31

2.32

Red and Green

The *Edible Boston* hot-pink title pops against a mint green table. Look closely at the photo and notice the appeal of the variety of greens and pink foods is tempered by soft browns in the toasts.

When the pale green matchbook cover is opened, red matches strike our attention.

Yellow and Purple

The design for the Natural High festival is a vibrant use of two sets of complementary colors and promises a fun weekend. Bright yellow typography is highly visible against soft purples in the photo of this spread.

Introduction to Adobe Illustrator

Overview of tools in Illustrator
Color Palette in Illustrator
Drawing Tools in Illustrator

The Illustrator Toolbar
Of Illustrator's top ten tools—those that do most of the work—the most useful is the Selection tool. This black arrow is used for selecting an entire object or multiple objects.

You'll use an additional nine tools frequently to create projects. The white arrow is the Direct Selection tool. It is used to select a particular point on an object to make precise adjustments to graphics. The Pen tool will initially be challenging, but fun and rewarding, once you've had a bit of practice. It is used for drawing. Note the tiny black triangle at the bottom of some toolbar squares, which indicates multiple versions of a particular tool. The Rectangle tool has important variations, including the Ellipse and Line tools. The Type tool generally works in the same way as most type tools in word-processing software. However, when you put your cursor over the Type tool, then press and hold the mouse button, you'll see variations of this tool. These include the Path Type tool for flowing text along any sort of line such as circles, waves, or an illustration (we'll cover these in Chapter 3). You will frequently use the Rotate and Scale tools. The Gradient tool fills shapes with variations of chosen colors. The Eyedropper tool is useful for identifying objects' specific hues. The Artboard tool allows you to customize the size of your design file. At the bottom of the toolbar, you'll see a magnifying glass—this is your Zoom tool.

Here's a useful list of keyboard shortcuts to save time while creating projects. Good news: you can do unlimited undos in Illustrator. Depending on how much memory is available, you can undo your steps, in reverse order, by pressing Command + Z. At any time, press and hold the Command key to temporarily switch to the Selection tool while using another tool.

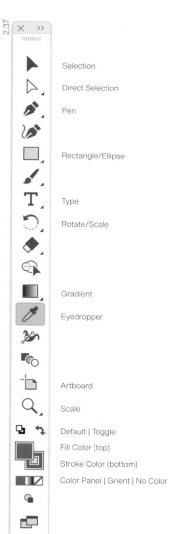

Selection
Direct Selection
Pen

Rectangle/Ellipse

Type

Rotate/Scale

Gradient
Eyedropper

Artboard

Scale

Default | Toggle
Fill Color (top)
Stroke Color (bottom)
Color Panel | Grient | No Color

Keyboard Shortcuts

See note in Chapter 1 about including PC commands here.

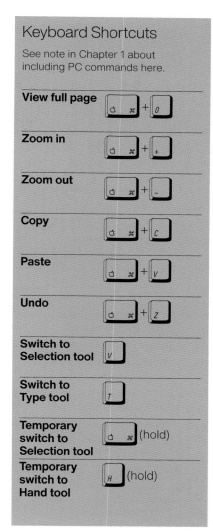

View full page	⌘ + 0
Zoom in	⌘ + +
Zoom out	⌘ + −
Copy	⌘ + C
Paste	⌘ + V
Undo	⌘ + Z
Switch to Selection tool	V
Switch to Type tool	T
Temporary switch to Selection tool	⌘ (hold)
Temporary switch to Hand tool	H (hold)

Tip

We work in the Essentials Workspace. It is the default setting for the software. You can also set it by choosing Window > Workspace > Essentials. If at some point you can't find the Properties panel or other window, simply reset the workspace by choosing Window > Workspace > Reset Essentials.

Color in Illustrator

Let's look closely at the color tools in Illustrator. The two large color squares in the toolbar are the Fill and Stroke colors. The color on top is the Fill color. The color below is the Stroke (outline) color. You can set either to "no color" by pressing the small button with a red diagonal line (below right). The bent arrow toggles between Fill and Stroke colors. Click on the tiny white and black overlapping squares at the top left, to swap to these customizable default colors. To choose new colors, double-click on the either the Fill or Stroke square to open the Color Picker window.

The Color Picker window is the best tool for selecting unique colors in Illustrator. At first glance it has an intimidating array of coded letters and numbers, but a bit of study will allow you to understand every aspect of this window.

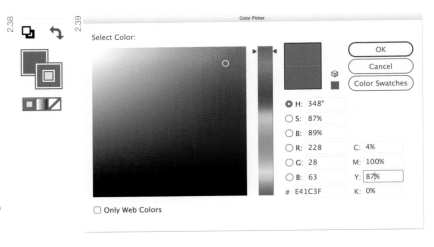

HSB	Name	Range	Effect
H	Hue	0 to 360	location on the standard color wheel
S	Saturation	0% to 100%	gray to fully saturated
B	Brightness	0% to 100%	black to white

Hue | Saturation | Brightness

In the Color Picker window, you see several letters. The first column contains HSB.

Hue, simply stated, is the name of the color. Illustrator displays a numeric value for hue that indicates the location of the color on the color wheel in degrees between 0 and 360. To understand this range, imagine the rainbow strip in the Color Picker is formed into a circle with red at the top. Think of the circle as a clock face. Near the top, or 12:00, is 0° for red. Yellow would be around 4:00, or 60°, and blue around 9:00, or 200°. Position the slider at yellow and blue on the hue spectrum and note the hue values.

Saturation can be thought of as the intensity of the color. Also referred to as chroma, it indicates the amount of gray in a color. Notice where the darker values of the color appear in the color field (the large square) inside the Color Picker

window. The higher gray-content colors appear in the bottom third of the square. These colors are referred to as muted or shades. Highly saturated colors contain less gray. Consequently, the most saturated colors may be found in the upper-right-hand corner of the color field. Highly saturated colors are referred to as bold or true. Brightness is the amount of white in a color. Brighter hues appear in the upper left of the Color Picker square and are often called tints.

Out of Gamut Sign

A tiny triangle with an exclamation point sometimes appears next to the current and previous color rectangles to indicate that the current color is out of gamut. Gamut is the range of colors in any given color space. Designers work with many devices (computer displays, inkjet, and laser printers); each has a different color space. For example, a laptop display shows more colors

than inkjet printers, but printers can produce purer blacks. When a color cannot be displayed on a device, it is out of gamut. Heed the warning when producing work for clients—if the sign appears, click the triangle and the selection will switch to the closest printable color.

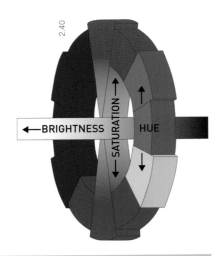

Initial	Color	Range
C	cyan	0% to 100%
M	magenta	0% to 100%
Y	yellow	0% to 100%
K	black	0% to 100%

Initial	Color	Range
R	red	0 (black) to 255 (white)
G	green	0 (black) to 255 (white)
B	blue	0 (black) to 255 (white)

CMYK
Cyan | Magenta | Yellow | Black

CMYK is a pigment-based subtractive color system that is synonymous with the four-color printing process. C is for cyan, a bright blue. M is for magenta, a bright pinkish-red. Y is for yellow, and K is for black. These four colors are used in combination to print all the colors in the spectrum.

This system has been used in various ink-based printing technologies for well over a hundred years: first lithography, then offset printing, and now desktop printing. When you use a color ink cartridge for your desktop printer, note the colors on the package: they are typically CMYK. The subtractive color system describes how we perceive ink-based color. As light strikes a printed page, the ink absorbs (subtracts) a portion of the color spectrum; what is not absorbed reflects back to the eye, resulting in the colors we see. When cyan, magenta and yellow are all absorbed by a pigment, the result is black, as the ink subtracts the colors.

RGB
Red | Green | Blue

The computer or phone display is a light source, so these primary colors are not the pigment-based primaries of red, yellow, and blue. They are instead the additive primary colors, red, green, and blue, also known as RGB.

Your computer screen creates color by illuminating red, green, and blue phosphors. When the colors combine, they can create most of the color spectrum. When all colors are combined at their fullest strength, they create white; consequently, this is termed the additive color system. To produce digital display graphics, designers work in the RGB color mode.

The RGB color mode provides brighter variations of colors compared with CMYK. When beginning a new project, always set the color mode initially. Choose the color mode by determining how the final design will be displayed. If the final design is to be printed, use the CMYK mode. If the final design will appear on a digital display, use the RGB mode. Avoid switching color modes within a file: your designs will lose color information.

The Pantone Matching System

Adobe provides designers the ability to view color swatches in libraries and to organize unique groups of colors into swatch libraries. The most widely used of these is the Pantone Matching System.

Pantone is an internationally used color system accessed through the swatch library. You should be aware of (not alarmed by) the term 'PMS.' PMS numbers refer to the Pantone Matching System. This is a standardized numbering system for colors, which enables designers and producers to specify particular hues by number. The use of PMS numbers guarantees that the specified hue will be reproduced consistently, no matter who or what is producing the work. For example, the Canadian government's maple leaf has a government-decreed Pantone number (PMS 032) so that the country's symbol is displayed consistently, whether on a government website or on a ferry boat to Vancouver Island.

When working with a particular company, designers specify the brand's PMS color. For the Tiffany's

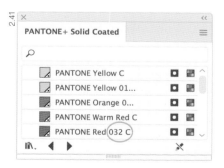

website, designers select the signature robin's-egg blue that has its own Pantone number. This particular PMS number, 1837 (the year the store was founded), is patented, and its use requires approval by Tiffany's and Pantone.

In Illustrator, to access the PMS color chart under the Window menu, choose Swatches; then from the Swatch window's top-right hamburger menu, choose Open Swatch Library > Color Books > find the Pantone listings, and select one that suits your project, this one is Pantone + Solid Coated.

The most common choices are Pantone Solid Coated or Pantone Solid Uncoated. These terms refer to the finish of the paper: coated paper is very smooth, so the ink appears brighter. Uncoated paper is more porous, so the ink appears somewhat less vibrant. Designers must specify whether the color is coated or uncoated when submitting a job to a printer. These colors are also referred to as spot colors.

Each Adobe software accesses Pantone listings differently. In Photoshop, double-click on the toolbar foreground color to open the Color Picker window, next click on the Color Libraries button, then choose from the Book pull-down list; a common choice is Pantone + Solid Coated. In InDesign, from the Windows menu, choose Color > Swatches, now from the top-right hamburger menu, choose, New Color Swatch; in the New Color Swatch window, choose Spot, as the Color Type, and Pantone from the Color Mode list. You'll get used to it.

Only Web Colors

Only Web Colors is no longer relevant. If you are interested in the history, keep reading; otherwise skip to the next paragraph. Designers adhered to a web-safe color palette when websites began proliferating in the nineties until roughly 2009. A web-safe color palette acknowledged that while your monitor may have been capable of displaying millions of colors, web browsers displayed only 216 colors consistently. Checking the Only Web Colors button in the lower left of the window allowed you to view only these 216 colors.

Significant to website design, a six-digit hexadecimal number is listed at the bottom of the left-hand column, indicated by the # box. Website software such as CSS defines colors using the hexadecimal (sixteen-character 0–F) numbering system. The hexadecimal value for white is FFFFFF, black is 000000, and one of my favorite blues to use for websites is 006699.

Now that you are expert in the use of the digital Color Picker, let's create a color wheel using drawing tools in Illustrator.

Design + Software Skills 2.1: Illustrator Drawing

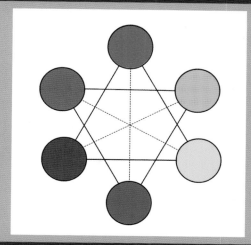

Objectives

This deceptively simple exercise achieves several objectives, allowing you to develop skills in software drawing and deepen your understanding of color theory. The repetition required to draw the color wheel reinforces the ability to use the new tools.

To Do

Draw the color wheel using Illustrator's drawing tools.

1 Make a new file

Start Illustrator and choose File > New or Press Create New. Select Print from the top line of the New Document window that appears. Select your preferred unit of measurement on the right pull-down menu. This will provide you with default letter size and set the color mode to CMYK. (If you choose Web or Mobile, your file will have pixel dimensions, and the color mode will be RGB.) Press Create. A blank document will appear on your screen. We're ready to go!

2 Confirm Essentials Workspace

We will use the Essentials Workspace throughout these Software Skills. Confirm you have the Essentials Workspace on the top right of your window. If not, go to the Windows menu and select Workspace > Essentials. If a panel accidentally closes, you can always refresh by selecting Windows > Workspace > Reset Essentials.

3 Choose a color
Select a red.

(a). First double-click on the Fill Color square

(b). To see a selection of reds in the large square, slide the arrows on the rainbow spectrum to the bottom

(c). If your Color Picker does not appear as in this image, confirm that H (hue) is selected

(d). Choose a true, saturated red by moving your cursor to the top right of the Color Picker square

Click OK to select red and close the window.

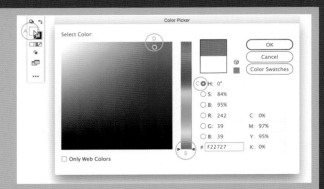

4 Make a circle
Place your cursor over the Rectangle tool, click, and hold. Select the Ellipse tool and release.

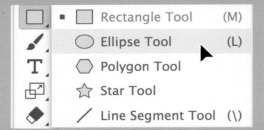

5
While the Ellipse tool shows in the toolbox, click once on your file workspace. Illustrator will place your object wherever you click, so click in the top center area of your page. Make the height and width 1 inch (2.5 cm), and then select OK. You've successfully drawn your first object. Admire its bold symmetry for a moment.

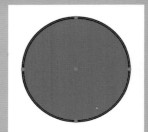

6 Make more circles

Now, draw the other colors in the wheel. Reposition your cursor lower and to the left to create another circle.

Next, double-click in the Fill Color square again to bring the Color Picker window into view. Move your cursor along the hue spectrum bar to choose a blue, then select a specific true blue from the large square color field.

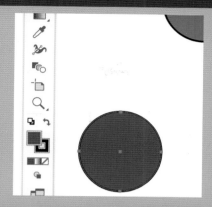

7 Make the wheel

Repeat the above directions to draw another circle and make it yellow.

If your virtual triangle looks a bit wobbly, you can move a circle after the color has been changed. Choose the black Selection tool, click over a circle, and while holding the mouse button down, nudge it into position.

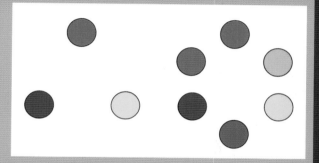

Tip

A shortcut to the Selection tool is to hold down the Command key. Your cursor will temporarily change to the Selection tool, allowing you to move the circle. Once the circle is moved, and you release the Command key, your cursor returns to the Ellipse tool mode.

Continue to follow the directions in steps 4–6 and add three more circles—orange, green, and purple—to complete the wheel. Position the colors into a wheel using the Selection tool.

8 Draw the lines

Draw black lines to connect the two triangles that indicate the primary-color and the secondary-color groups.

Select the Rectangle tool and hold to select the Line tool from the sub-tool window. The Line tool will automatically switch your color selection to black stroke and no fill color.

Place your cursor on the lower-left-hand curve of the red circle, hold the mouse button down, and drag the cursor to the top right of the blue circle, then release the mouse button.

Repeat this step to draw solid black lines among the color groups.

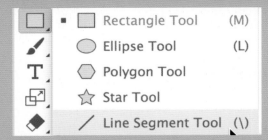

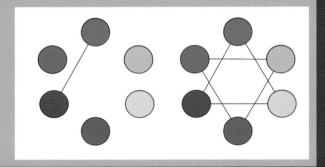

9 Make adjustments

If you want your lines to meet perfectly with the black circles, make small adjustments to one end of a line with the Direct Selection tool. Choose the white Direct Selection tool.

First, "click off" by clicking into the white background area so that nothing is selected. Then click directly on the end of the line you wish to adjust. Click on the blue square at the end of the line and drag it into position.

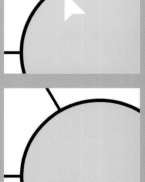

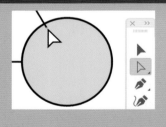

10 Draw dashed lines

Draw three new lines between the complementary color pairs (red and green, blue and orange, purple and yellow). These three lines will cross in the middle. Use the black Selection tool to select all three lines by dragging a little rectangle at the intersection of these three lines.

From the Windows menu, open the Stroke palette. Check the Dashed Line box and fill in values for the dashes and gaps. Start with 2 pts each, then try experimenting with other values.

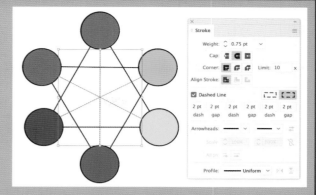

11 Finishing off

You can make the lines and circle outlines thicker by selecting them and then increasing their weight in the Stroke window.

Save your work by using the keyboard shortcut Command + S. Phew. You have completed your first Illustrator drawing. Well done. Break time!

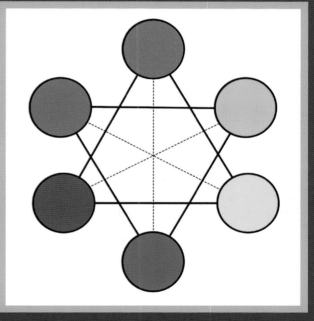

Design + Software Skills 2.2:
Photoshop Vector vs. Raster Images

Objectives

Demonstrate the differences between Illustrator and Photoshop image types, called vector and raster images. This exercise also shows the compatibility of Illustrator and Photoshop. In addition, you'll use one of the fun features of Photoshop: Filters.

To Do

Copy the color wheel from Illustrator to Photoshop. Compare the two types of images: vector and raster. Then get creative in Photoshop by applying various filters to the color wheel graphic.

Photoshop was introduced in Chapter 1. Now let's expand your Photoshop skills.

1

Create a new file

Open Photoshop and create a new file by choosing File > New. Choose Letter size, inches, Resolution set to 300, and leave the remaining options as the default values. Press Create and a new file will open on your screen.

2

Copy the wheel

Return to your Illustrator color wheel file for a moment to copy it into Photoshop. In Illustrator, click on the file. Use keyboard shortcuts to select the entire image and make a copy. Press Command + A, then Command + C. Your image will look like this when the color wheel is selected.

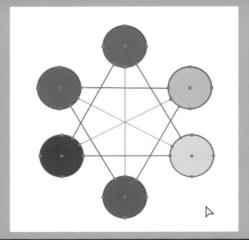

3

Paste in Photoshop

Return to your Photoshop file by clicking
on the Photoshop icon in the dock. Press
Command + V to paste the color wheel into
your file. Photoshop will prompt you for a
Paste type. Select Pixels, because this graphic
will be converted to a raster image. Then click
the OK key to place the graphic into your file.

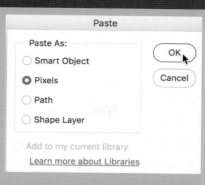

4

Layers

Note that the graphic is placed into a new
layer. Remember, Photoshop creates a
new layer every time you paste an object
into a file. Layers allow you to edit graphics
independently; it's a very good thing.

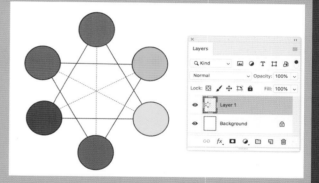

5 Zoom in

Use the Zoom tool to magnify the graphic. Select the Zoom tool from the toolbox. Click and hold the mouse button down as you drag to draw a virtual rectangle around the area you wish to magnify. Release the mouse button.

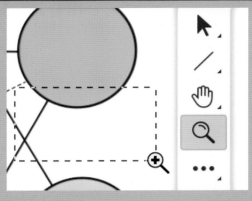

Raster images

Notice the rather rough-looking edges of the magnified illustration. Photoshop files are raster (or bitmap) images. These are a collection of pixels in a dense grid. Pixels look like tiny squares. A raster image can display subtle variations in colors, as is typical of photographs.

But raster images cannot be enlarged without causing problems because a set number of pixels is based on the image resolution. When an image is resized, the number of pixels remains the same. Consequently, as the size of the pixel enlarges, the image acquires a blurry appearance. Designs created in Photoshop are raster images.

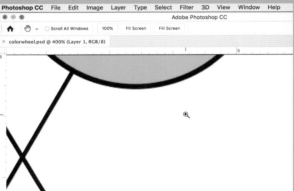

Vector graphics

Here is the same graphic magnified in Illustrator. Notice how smooth the image appears in Illustrator. Illustrator creates vector graphics. This term refers to the mathematical interpretation and display of graphic objects.

Vector graphics can be resized without losing quality. The crisp lines of vector graphics are highly desirable, particularly in print and logo design, because they never lose their details or clarity. Designs created in Illustrator are vector graphics.

Rasterization is the process that occurs when a vector image is converted into pixels.

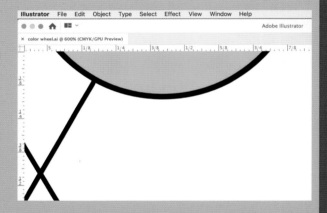

6 Zoom out

In Photoshop, return to the full-size image. Press Command + 0 (zero).

Photoshop Filters

Let's have some fun with Photoshop. The filter menu provides you with countless variations for modifying graphics. While many of the filters may not be professionally viable, they'll provide you with an early opportunity to get creative using the software.

1 Apply a filter

While the color wheel is selected, choose the Filter menu > Pixelate > Color Halftone. Use the default values initially and press OK.

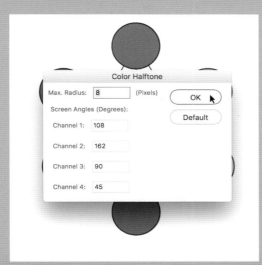

2 Experiment

Now experiment with these settings and other filters. What happens when you change the Max. Radius or other settings?

Illustrator also has filters; they can be found under the Effects menu. Experiment with the options and notice how the effects are similar or differ from the Photoshop results.

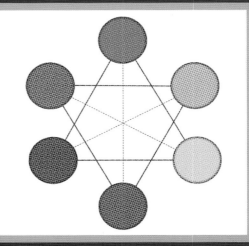

Design Analysis:
Complementary Color Grids

Objective

Provide a reinforcement of the color theory discussion and develop the ability to see nuances among colors in printed material. In the next Illustrator exercise, we'll use this hue analysis project to learn how to select colors in Illustrator.

To Do

Create two grids for a pair of complementary colors.

1 Search printed material, such as magazines or catalogs, for examples of one primary and one secondary color of a complementary color pair.

2 Cut out nine 1-inch (2.5 cm) squares of each color.

3 Arrange each color's squares on a 3 x 3 grid.

4 Place the most "true" of each color into the center of each grid and arrange the other squares around it.

5 Paste the grids into your notebook.

My student Val's homework is shown here. She used blue and orange, but you can choose any of the complementary color pairs to create your hue analysis.

2.42

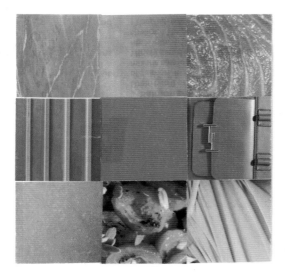
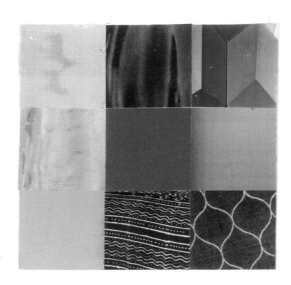

Design + Software Skills 2.3: Illustrator Color and Drawing

Objectives
Provides further reinforcement of color theory and develops the eye to see nuances among colors on the computer screen. Expands the ability to interpret CMYK values and to modify them in the software. Develops facility with Illustrator while reinforcing the use of tools demonstrated in this chapter's first exercise.

To Do
Recreate the hue analysis color grids in Illustrator using digital color tools. Use the Color Picker window options to replicate the printed colors most closely. Once the grids are completed, create pairs of colors from both grids. Explore contrasting and harmonious color combinations. These color pairs will be used in the next exercise.

1 Make a square

Start Illustrator and open a new document that is letter size and CMYK. Refer to the color wheel in Software Skills 2.1 for a reminder about the Illustrator New Document window.

Choose the Rectangle Shape tool and click once on the file towards the top left of your page. Make the width and height 1 inch (2.5 cm). Make the stroke no Fill by clicking on the lower, Stroke color square, then click on the red-diagonal-line button below.

Look closely at the results of your hue analysis exercise. Now double-click on Fill Color in the toolbox to open the Color Picker window. Then you'll choose a hue that most closely resembles the top-left-hand color in your paper grid.

Click OK and a square with the same color as the top-left-hand block in your grid will appear in your file.

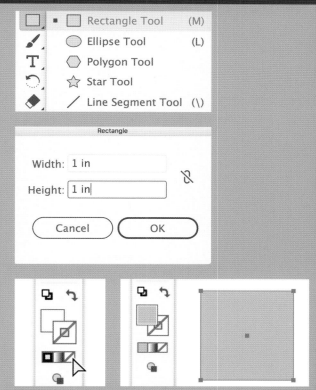

2 Make a row

Create the entire grid by copying and pasting this one square. First, make the top row. Choose the black Selection tool and click on the square. Press Command + C, then press Command + V to paste. Once again, press Command + V to paste the third square.

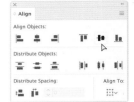
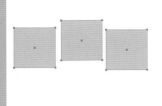

Make the squares line up neatly. Click slightly above the top left of the first square, hold the mouse button down, and drag to the right over all three squares. Release the mouse button. You have selected all three squares. In the Properties panel, press the Vertical Align Center icon. While the squares are still selected, distribute your squares evenly. Press the Horizontal Distribute Center button.

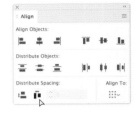
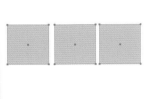

3 Build the grid

While all three squares are selected, copy and paste the entire row. Press Command + C, then Command + V to paste. Move the second row of squares carefully into position.

Press Command + V to paste the third row. Move the third row into position.

You have drawn a perfect grid.

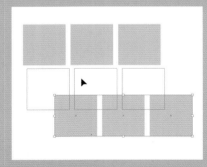
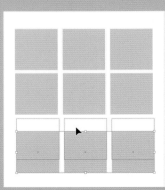

4 Select colors

Now change the other eight squares to look like the colors in your printed version. Select the second square to change its color.

While the second square is selected, double-click on the Fill Color to open the Color Picker window. Move the arrow on the vertical rainbow spectrum towards blue, then choose a color from the large area. Notice the HSB, CMYK, and RGB values as you go through the grid.

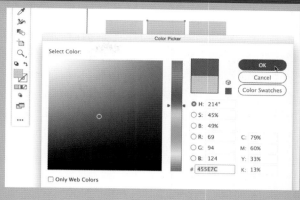

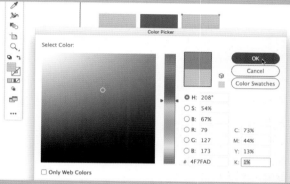

5 Adjust the colors

Select each square one by one and find the color that most closely resembles your paper squares. Remember to double-click on the Fill Color in the toolbox to open the Color Picker window.

As you choose each color, notice the position of the cursor and the corresponding HSB, RGB, and CMYK values. Can you fine-tune the color selection by changing the numeric values?

6

Draw a guide

You can draw guides to help you place the next grid. Press Command + R to show rulers along the left and top edges of the file. Then click in the top ruler, hold your mouse button down, and drag a guide into position.

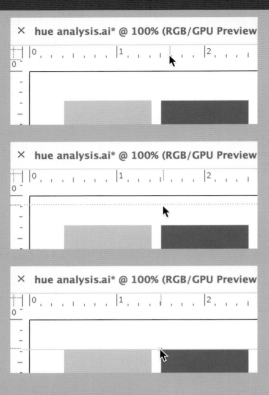

7

Create the second grid

Repeat this hue analysis process with the other color grid. Select the entire first grid, press Command + C, then Command + P to copy a second grid.

While still selected, move the second grid to line up with the guides you've drawn. Then change the colors to duplicate the printed version.

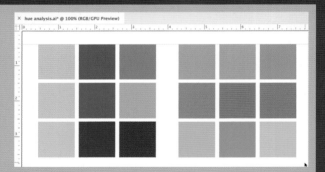

8 Complementary color pairs

Using the grids created in the hue analysis file, you can choose effective combinations of complementary color pairs. Reposition the second grid below the first on the page. Using the black Selection tool, drag around the entire second grid and, while all nine squares are selected, position them below the first grid. Use the black Selection tool to click on one color from your first grid. To copy, press Command + C, then Command + V to paste.

Hold the mouse button down and drag the square to the top right of the page.

Now, select a square from the second grid that works well with the square on the right. To copy, press Command + C, and Command + V to paste. Select and hold the mouse button down to drag the second square to create a pair.

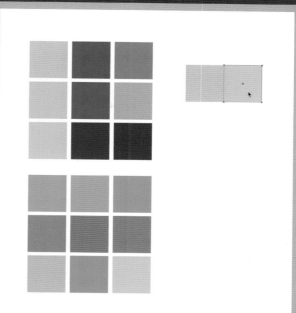

9 Finishing off

Create four pairs of complementary colors.

Tip

**Use guides to help create this chart. You may choose to use Smart Guides as you gain experience in Illustrator. Make Smart Guides visible by choosing View > Smart Guides. There are additional guide options under the View > Guides menu.
Guides do not print. Pressing Command + ; will make them temporarily invisible.**

Save and print this file using the File menu, or press Command + S and Command + P.

You can use these colors for additional projects.

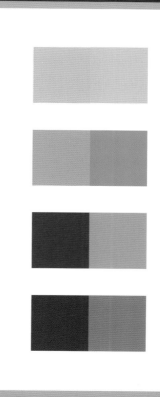

Major Points Summary

— The elements of design are color, direction, line, shape, size, texture, and value.

— Designers manipulate one or more elements to create designs.

— Elements are the same no matter what the media, whether designing a logo or a website.

— Lines are so varied, they are described using three terms: type, direction, and quality.

— Shapes can be tangible or abstract, flat or dimensional, and filled with color, texture, pattern, or value.

— Value is the relative lightness or darkness of a color.

— Contrast describes the relative difference of an element as it is used in a design.

— Contrast provides visual interest in designs.

— High contrast means a lot of difference, while low contrast suggests minimal differentiation.

— High contrast is used functionally: for legibility (black letters on white paper) and for hierarchy of information (larger text leads the viewer in and provides titles, while smaller text provides the details or the story).

— Texture can be purely visual (photographic) or tactile (actual).

— Size can be used to introduce contrast and hierarchy.

— Objects placed in unusual directions can be used to attract attention in a layout.

— Unity is a basic principle of design; all objects on a page should be visually harmonious.

— Color has an immediate effect on the audience. Even before copy is read, color makes an impression.

— Greens and blues are cool and soothing.

— Yellows and reds suggest warmth, even heat.

— Choose colors that convey meaning and are appropriate for your client; avoid using only your favorites.

— Primary colors are red, blue, and yellow.

— Secondary colors are green, orange, and purple.

— Complementary color pairs are red/green, blue/orange, and yellow/purple.

— Complementary color pairs appear opposite each other on the color wheel, and therefore have the most contrast. This makes them visually dynamic when used together.

— CMYK (cyan, magenta, yellow, black) is the four-color process system for printing.

— CMYK represents the pigment-based subtractive color system. Combine the primary colors, red, blue, and yellow, to create the rainbow spectrum. As light hits a printed page, ink absorbs (subtracts) a portion of the color system; we see what is not absorbed when the light reflects back.

— CMYK can produce pure blacks. Choose this mode for projects to be printed.

— RGB (red, green, and blue), the primary colors of the additive color system, are used in displays such as TVs and computer displays.

— Your computer displays colors by illuminating red, green, and blue phosphors. They combine to create most of the color spectrum. When all the colors are combined (additive) they create white.

— RGB produces brighter colors than CMYK. Choose the RGB mode for projects to be displayed on screens such as laptops and phones.

— Pantone is a color standardization system used internationally, from software to print shops, for consistent appearance of spot colors.

— Website software identifies colors with six-digit hexadecimal numbers.

— Vector images are created in Illustrator and can be resized with smooth edges.

— Raster images are grids of tiny colored squares that combine to form an image.

— Rasterization occurs when vector images are converted into pixels.

Software Skills Summary

— Illustrator introduction.

— Overview of the toolbar and the most commonly used tools for beginning designers: Selection, Direct Selection, Pen, Type, Path Type, Vertical Type, Line, Rectangle, Ellipse, Rotate, Resize, Zoom, Fill Color, Stroke Color, Toggle Color Arrow, Default Color.

— Keyboard shortcuts.

— Skills: Color Picker window, color terminology and values, shapes, stroke, dashed lines, align options, guides, scale proportionately, group and ungroup, type, create outlines, rotate.

— Color terminology: hue, saturation, brightness, CMYK, RGB, out of gamut, web-only, hexadecimal numbers, Pantone.

— Tools covered in depth: Ellipse, Selection, Line, Direct Selection, Zoom, Rectangle.

— Transfer files across software, Photoshop filters.

— Digital-image terms: bitmap images, raster images, vector images, rasterization.

Recommended Readings

More than any other topic, I read about color. Some of my favourite books include the following:

Joann Eckstut and Arielle Eckstut's *The Secret Language of Color* explores color in science, nature, history, and culture.

The history of seventy-five colors is covered in *The Secret Lives of Color* by Kassia St. Clair.

And, for smiles, find out what your birthday color means in Michele Bernhardt's *Colorstrology*.

The Complete Color Harmony: Pantone Edition by Leatrice Eiseman. Eisman is the executive director of the Pantone Color Institute, and she annually leads the team that selects the Pantone color of the year. This book gives the insider scoop on that process and is a good resource on both color theory and smart Pantone color combinations.

Radiolab Podcast: Why Isn't the Sky Blue? Produced by Tim Howard, WNYC Studios: https://www.wnycstudios.org/podcasts/radiolab/segments/211213-sky-isnt-blue

Typography

Paula Scher, *Bring in 'Da Noise, Bring in 'Da Funk*, The Public Theater poster, 1995

Paula Scher has said, "Typography is painting with words. It's my biggest high." Using uneven letter spacing and a variety of weights, her groundbreaking design visually expressed the vibrant music and dance of Savion Glover's tap-funk show. The poster and logo's different sizes, weights, and placement represent the great variety of people in New York City. "It's crazy, it's in your face, it's New York," says Scher of this beginning of a three-decade collaboration with the Public. The active and unconventional visual system was designed as an expression of the theater's mission to provide a venue for innovative performances. Variations of this typographic style began appearing everywhere, from advertising to logos. Everyone stole the design, so later Scher switched to serif fonts just to be different.

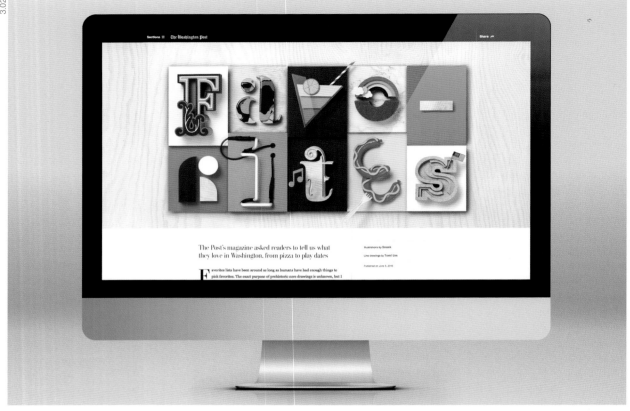

Identifying Typefaces and Making Hard Choices

This screen is for the *Washington Post*'s first Readers' Choice Awards for best Eating, Drinking, Music and Arts. The designers handmade each letter using materials that reflect the essence of each category. For example, the *V* looks like a fancy cocktail and was made with glass.

A fascination with typefaces often leads students to careers in graphic design. Others feel overwhelmed by too many font choices; with so many decisions to make about style, size, and placement, they become intimidated by typography. This chapter will address both attitudes. Everyone can benefit from this introduction to terminology and strategies for selecting and using type. Type geeks will be inspired by the

opportunities it offers to use letters in new ways. Those with type-phobia will find easy-to-use tips, along with engaging exercises and projects.

Terminology

The illustration below shows key terms that aid description, your ability to combine typefaces effectively and design layout. The terms labelled within the gray lines are essential for novices.

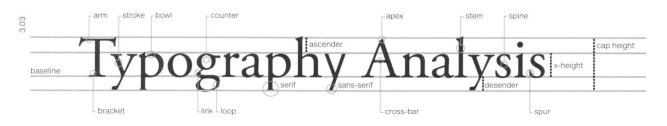

Garamond | "Good communication is as stimulating as black coffee and just as hard to sleep after." *Annee Morrow Lindberg*

Is It a Typeface or a Font?

Do you use the terms typeface and font interchangeably? If you do, you're not alone, but actually there is a distinction between them—admittedly a fuzzy one—with regard to digital formats. Typeface refers to the overall character set of a particular design. Font refers to the complete set of characters in any given size and style. Throughout printing history, typefaces have been designed and fonts are their product. For consistency, I use font to refer to the software version and files; I use typeface when referring to the particular design of the letters.

Serif and Sans Serif

Serifs are the notches at the ends of letterforms. The broadest categorization of typefaces can be determined by the characteristics of the serifs. All serif typefaces have distinctive notches at their terminals. A typeface without serifs is called sans serif (*sans* means "without" in French).

Serif typefaces are based on type forms developed in the first century AD particularly for Roman monuments. These imposing stone structures recounted battles by listing names, dates, and other details. The engravings, carved with chisels, were very stately and enduring, in a type style called Monumental Capitals (*capitalis monumentalis*). Historians believe stonemasons used chisels at the end of the letterforms to create a tidy appearance to the inscriptions; this may be the origin of serifs. Perhaps the best-known example is Trajan's Column, c. 113 AD. The original can be seen in Rome, and a replica is in London's V&A Museum. A popular font based on this inscription is suitably named Trajan. Given serif typefaces have been in use for 2,000 years, they are considered to be conservative in style. They are also highly legible when used with large quantities of type, as they guide the eye along the line of text more quickly. A popular example of a serif font is Times.

Type Types: Serif Style Categories

Serif typefaces can be categorized into Old Style, Transitional, Modern, and Slab Serif. These groups provide designers with strategies for selecting and combining typefaces. Many of today's popular typefaces were developed hundreds of years ago by innovators whom we know now as names of fonts.

Old Style

Garamond is an Old Style typeface. Old Style refers to typefaces with characteristics that were typical when metal type printing was adopted in Europe. These letters have low contrast between the thick and thin strokes, with gradual blending to slanted serifs. This form is indicative of two factors: the typeface was modified from a handwritten style, and the existing technology would need further refinements to allow for more delicate shapes. The developer of this typeface, Claude Garamond, was a renegade. At the time of Gutenberg's invention of moveable type in mid-fifteenth-century Germany, books were written, designed, published, and sold in the same shop. In France, eighty years later, Garamond became the first type designer to work independently of a print shop. His carefully drawn and proportioned fonts were influential in replacing the Gothic handwritten style of text. Wildly successful, Garamond's typefaces became the dominant style throughout Europe and remained so for over a hundred years. Use Garamond, Palatino, and other Old Style typefaces when your page design needs to achieve an overall uniform appearance. Do you have a lot of text that you want people to read? Use an Old Style or a Transitional typeface.

Transitional

Transitional typefaces have more refined forms than Old Style ones, with more contrast and deeper bracketing to the serifs. Technical developments by the mid-eighteenth century included smoother inks, glossy papers, and stronger metal alloys to support these elegant designs.

In Birmingham, England, the wealthy young entrepreneur John Baskerville, who made a fortune inventing varnish, turned his attention to printing technology. Baskerville developed more delicate typefaces that could withstand repeated poundings of the press. He also produced/devised smoother paper and inks so that these finer typefaces could print without breaks or clogs. His designs had an airy quality because of the lightness of the letterforms and the generosity of the page margins. Ever the perfectionist, Baskerville melted down his type after each printing for consistent pristine results.

Times, another Transitional typeface, is perhaps the most widely used serif font in English-speaking countries. The characteristics of Transitional typefaces are not readily noticeable—they do not attract our attention. Nothing about their design distracts or hinders our reading; consequently, they are highly reliable and legible. Use Transitional typefaces for readability and subtle elegance.

Modern

The Modern letter style is characterized by a distinct contrast between the thick and thin strokes and hairline serifs abruptly set at right angles to the stems. Giambattista Bodoni's typefaces are in the Modern category because the Italian used the latest technology developed during the Industrial Revolution for the design and implementation of his work.

Celebrated for his accomplishments, Bodoni eventually developed 300 typefaces. Years after his death, Bodoni's hometown of Saluzzo, Italy, erected a statue in his honour—pause for dreadful irony—with an inscription carved in Old Style letters.

Use Modern typefaces for their cool, crisp appearance. Brands from Vogue to The Gap use them in their logos and mastheads. This style is not the best choice for large amounts of copy—the extreme contrast between the thick and thin strokes creates an overall irregular appearance on the page.

Slab Serif

In the mid-nineteenth century, before images could be reproduced easily, advertisements used bold typefaces to grab attention. Stout typefaces were designed with serifs that were the same weight as the vertical and horizontal strokes. At the time this style was developed, excavations and artefacts from Egypt were quite the rage. Consequently, this category was originally called Egyptian. Fonts were given Egyptian names, such as Memphis, to increase their sales.

Common examples today are Rockwell and Courier (the typewriter font). Clarendon and New Century Schoolbook are Slab Serifs with great legibility. With thicker strokes and serifs, they create an overall darker appearance to the page. Many children's books use these bold, clear typefaces to ease the new reader's experience. In your page layouts, try Slab Serif titles combined with sans-serif body copy for a clean yet dynamic page appearance. Slab Serif fonts are wonderful to use for digital products, such as apps and websites, because Slab Serif fonts are mono-weight— the thickness of each letter is consistent, and this is beneficial for digital displays.

3.05 **Baskerville** | "What I cannot love, I overlook. Is that true friendship?" *Anaïs Nin*

3.06 **Bodoni** | "It always seems impossible until it's done." *Nelson Mandela*

3.07 **Rockwell** | "Treat your friends as your pictures and place them in their best light." *Jennie Jerome Churchill*

"There is no charm equal to the tenderness of heart." Jane Austen

Beyond Serifs

Sans Serif

Just like fashion trends, typeface designs evolve. By the late nineteenth century, Slab Serif typefaces were predominant. The next development was to chop off the serifs. Also called Gothic, such typefaces were popularized by the Bauhaus designers in the 1920s. Sans-serif typefaces exemplified the school's anti-ornamental approach to design. Helvetica is a sans-serif font, considered by many to be a perfect font. It is so popular that a movie was made about it: *Helvetica*.

Sans-serif typefaces have a more casual style than serif typefaces, yet they are usually just as legible. Sans-serif typefaces, such as Myriad, are frequently used for online text. Websites, email, and text messaging are often produced with sans-serif typefaces because their mono-weight lines result in better legibility on computer screens. Mono-weight means there is no variation in the width of the strokes and stems of the letterforms. Avenir is a good choice; it is a versatile sans-serif font that is used for signage at the airports in Hong Kong and Dallas, as well as by the Girl Scouts of America, the city of Amsterdam, and LG Electronics.

Though they are mono-weight, and have neither serifs nor subcategories, typographers have nonetheless developed distinctive sans-serif designs that exude style.

Three early twentieth-century typefaces, Futura, Gill Sans, and Franklin Gothic, continue to look contemporary today and provide specific tones to your designs.

Futura letters, with their spiked apexes and diagonal ends, are dynamic—wonderful in logos and titles. Gill Sans letters are relatively tall and elegant, with wide-open negative spaces, providing an easy-to-read and sophisticated appearance—great for large amounts of copy. Franklin Gothic letters are stout. The typeface seems a bit stern—use it for No Parking signs.

Today's type technology developments focus on the use of computer screens for type display. Typographer Matthew Carter currently works in Cambridge, Massachusetts but began his study of the field in Cambridge, England. Carter's career spans from metal type to digital fonts. He honed his skills in Holland, worked in London, then moved to Massachusetts where he co-founded Bitstream, an early provider of computer fonts.

Carter developed dozens of typefaces. Many are sans serifs designed specifically to improve legibility. These include the ubiquitous Verdana, originally created for Microsoft, and Skia, which is based on ancient Greek letters and was commissioned by Apple. The fonts commissioned by Apple and Microsoft ease reading on computer screens. Their x-heights are taller and counters larger—these features facilitate browsing text on websites. Verdana, a sans-serif design, has a singular capital letter I—it has serifs (!) to distinguish it visually from the number one.

Roger Fawcett-Tang designed this book. His work on my first book was a key factor for its success. Here are Roger's thoughts on the font he chose for this edition: "When Monotype released a totally re-engineered version of Helvetica 'Helvetica Now' in 2019, I bought a copy straight away! Neue Helvetica was always a popular font, but the kerning was often problematic and it lacked the refinement of other fonts. I often used Unica, which is a hybrid of Univers and Helvetica, as I felt it had a better color and balance on the page. *Graphic Design Essentials* seemed like the prefect project to use Helvetica Now on, the font comes in three different versions— Micro, Text, and Display—different versions work really well within their appropriate sizes. I used the Display weights for the headlines, and Text weights for the main body of the book. The result gives a clean look on the page, and I think the large numerals on the chapter openers look stunning!" We all thank Roger for this elegant and easy-to-read design.

FUTURA MEDIUM

GILL SANS LIGHT

FRANKLIN GOTHIC

Scripts

Scripts appear to be hand-drawn, often suggesting calligraphy. Commonly used fonts include the elegant Snell, sassy Zapfino, heavier Viktor, and friendly Renata. Adobe Fonts has many options available to download when you have Creative Cloud. If your project would benefit from a font that mimics the rough edges of signatures, try Adorn Garland. Dafont.com is a wonderful resource for free fonts for personal and academic use. If used for a client, there may be low fees.

Tips for scripts: Avoid using all capital letters with scripts, and limit their use to small amounts of text, not entire paragraphs.

Sign Painter was created in an independent type foundry, House Industries, which thrives in otherwise sleepy Delaware. House fonts are rich with nostalgic details, yet they are hip and undeniably beautiful. Commitments to traditional techniques and a mastery of digital media are apparent in the details. For example, the Sign Painter font features many connecting ligatures that give them a distinctive hand-lettered flair, and their software makes contextual substitutions. The House Industries website sells fonts in economically priced sets and creates custom typographic projects.

This spread shows some of House's script fonts. On the left, a font with elaborate alternative initial forms can be used for capitalized titles in place of the defaults that work better as lowercase text. The right page shows fonts that reflect cultural influences on scripts, such as French, Spanish and Austrian.

For Agent Provocateur, House—with trademark wit—drew a calligraphic corset to enclose the signature script initials of the lingerie company.

Novelty

Want your design to express a specific attitude immediately? Use one of the thousands of novelty typefaces for the title. These fonts are the wild and wonderful eccentrics of typography, designed to express specific styles, personalities, or trends.

While they add a distinctive, immediate voice to your designs, many novelty letters can be difficult to read. With their unique shapes and embellishments, these fonts are not necessarily concerned with legibility. The trick to using novelty

typefaces, such as Giddyup, is to limit their use to titles or short amounts of copy. Never use them for a whole paragraph or more of text—your audience doesn't want to struggle to recognize letters while reading. My advice regarding novelty typefaces is similar to what you'd tell a friend about to apply body glitter: a little goes a long way.

You will quickly learn to distinguish typefaces in these general categories. The ability to recognize characteristics of typefaces is the first step in determining how to select and use them effectively.

Neville Brody digitally modified Helvetica letterforms in Photoshop to create the font FF Blur. Rather than simply providing information, fonts can also provide immediate emotional expression. Believing that design colors the content, Brody became a celebrated art director for magazines, for which he created hand-drawn titles and innovative layouts to enhance the experience of reading each story. His postmodern designs pushed the boundaries of legibility. He designed many distinctive fonts, including Arcadia, Insignia and Dirty. Brody was the art director for a major redesign of *The Times* of London in 2006, where he created a bespoke font to be used only by the newspaper: the renowned Times Modern.

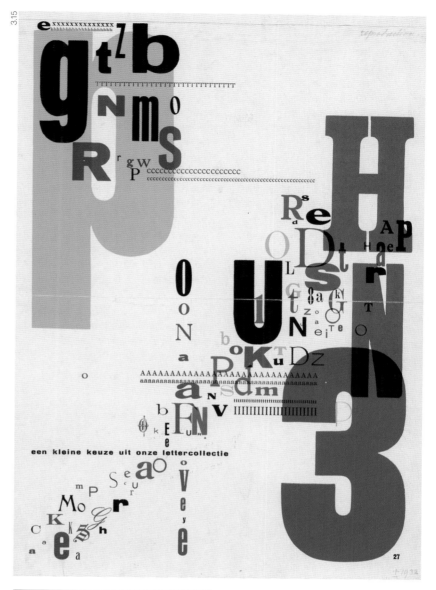

3.15

Piet Zwart, Drukkerij Trio (Trio Printers) catalog, 1931

Piet Zwart called himself a 'typotekt'. Originally an architect, he built graphic designs using type. Zwart, who was Dutch, was influenced by Dada, Constructivism and De Stijl. He was concerned for the reader and legibility, so he made his designs bold to attract attention. Zwart typically used brief slogans to allow the audience to get the message quickly.

He rarely used images; instead he relied on collage techniques, rhythmically placing letters of contrasting sizes and weights to create timeless appeal. This poster uses 134 different sizes and fonts. According to classic typography guidelines this mix really shouldn't work, but it does.

3.14 **Blur** | "Design is more than just a few tricks to the eye. It's a few tricks to the brain." *Neville Brody*

Choosing and Using Typefaces

There are so many fonts on my computer that my hand cramps as I scroll through the entire list. Does this stop me from searching for the perfect font? No. Next, I search Adobe Fonts and other favourite font websites and look there too. With an overwhelming number of fonts at our fingertips, how do we select the perfect one?

Start with a decision about the attitude and tone of your design. Then search for one or two typefaces that match the desired mood. Details, weight, and height create character. Heavy and bold suggests immediacy, while serifs can feel conservative, and thin serifs seem classical. Sans serif fonts often convey more contemporary or casual tenors. The right typeface sets the proper tone for your design and helps establish a voice for your copy.

Combining Typefaces— Opposites Attract

A classic approach to selecting typefaces is to use two fonts in each design: one for titles and one for the body copy—one serif, one sans serif. This mix adds visual interest to the page by creating contrast. It also enhances comprehension by indicating visual hierarchy of the text in titles and subtitles.

Why not combine two serif typefaces or two sans-serif typefaces on a page? Because two fonts from the same category will look somewhat similar—as well as somewhat dissimilar. **It frequently looks like a mistake when two typefaces of the same category are mixed.** See what I mean?

Don't distract your readers from the message by making this rookie mistake.

The coolest-sounding type term is also the most useful when combining typefaces: x-height. This is the height of the lowercase letter x. Try to use typefaces that have the same x-heights when combining two typefaces. In this way your design will achieve an elegant unity. A few winning type combinations of serif and sans-serif fonts are Didot with Avenir; Georgia and Helvetica Neue, and Benton Sans and Guardian. This blue chart is part of the branding manual for American Express. Can you identify the two categories of fonts?

Depending on the project, one of the two typefaces may also be a novelty typeface. The purpose of these is similar to the use of colors:

BENTON SANS
AAAAAA**AAA**
MMMMM**MMM**

Guardian
eeeee**eeee**
XXXXX**XXXX**

a novelty typeface communicates information before the words are even read. Use the distinctive typeface for the titles, perhaps the subtitles, and sometimes the folios (page numbers). The other typeface will be a serif or sans-serif font chosen for legibility. After all, the goal of most graphic design is to entice an audience and enhance its reading experience.

Most professional typefaces come in a variety of weights and sizes. A type family contains all the variations of a particular typeface. This chart shows Avenir in a series of variations described as light, light oblique, book, book oblique, medium, medium italic, black, and black oblique. These can be combined into a unified design because of the similarities in character widths, serifs, and x-heights.

Superfamily typefaces contain both serif and sans-serif versions. This greatly facilitates combining typefaces—use just one. These fonts include Meta/Meta Serif and Stone, which was developed specifically for new designers learning to combine typefaces.

Lou Dugdale designed this book's cover. She wanted a bold, modern effect, and experimented with many fonts and layouts. Lou ultimately chose the font Poppins for its strong form and readability. I love it! What do you think?

Avenir Light

Avenir Light Oblique

Avenir Book

Avenir Book Oblique

Avenir Medium

Avenir Medium Oblique

Avenir Black

Avenir Black Oblique

Distortions

Computer software has the ability to make changes to these carefully designed letters. Software allows type to be scaled and skewed with varying horizontal and vertical effects that do not maintain the original proportions of the letters. New designers need to respect the subtle, complex beauty of letter designs. Just because the software makes distortions possible, it doesn't mean we should indulge. Otherwise, the integrity of their forms is lost. Resolve to not modify the proportions of the original typography.

Fonts to Shun

Certain fonts are best unused— pass up poorly drawn typefaces with unsteady proportions. Avoid overused novelty typefaces, as their ubiquity weakens their appeal. Instead of the lovely but overused Copperplate, try classics like Baskerville or Garamond with small capitals. Typefaces that are trendy can make your design look dated quickly: consider the life span of your project when choosing the typeface. Every designer has a dislike for a particular typeface— ask around and heed the warnings.

Formats

The format of paragraphs affects the overall design as well as the readability of the text.

Justified type refers to lines of text in which all the lines are the same width. The left and right margins are virtually straight edges—gorgeous. This is the common format for books and magazines. It is considered the most legible format, good for reading with speed and ease—as long as the columns are not too narrow or too wide. A general guideline is to make your columns forty-five to seventy-five characters wide. Counting letters and spaces, sixty-six characters is considered ideal. For multiple columns, use forty to fifty characters. Justified copy is considered easier to read than unjustified copy, as the eye always knows where the line begins and ends. It also gives an orderly, elegant appearance to the page.

Unjustified copy can be flush left, right, centered, or asymmetrical.

3.18

3.19

3.20

We are sincere, fair and forthright, treating others with dignity and respecting their individual differences, feelings and contributions.

Left alignment is very commonly used in websites, magazines, annual reports, packaging, and captions. This format has a more informal appearance than justified copy. It is also called flush left and ragged right. It's the second most legible format, and it has very good readability because the eye always knows where the next line will begin. It cuts down on the use of hyphenated words. If you need to hyphenate, keep it to a minimum— never on consecutive lines and separate in accordance with the syllables.

Right alignment aligns the text along the right axis. It is most unusual in paragraph form because it is difficult for the eye to keep searching for the beginning of each line. It is best reserved for titles or a group of lines that help to balance a page layout. Here is a book cover with a right-aligned author and title. Right alignment is most commonly used to balance a design. See the wayfinding sign below: the top two destinations are left aligned and the bottom is right aligned. This letterhead design left aligns the contents and right aligns the logo and bottom address in blue for a smart, classic design.

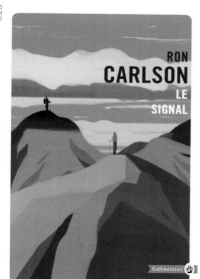

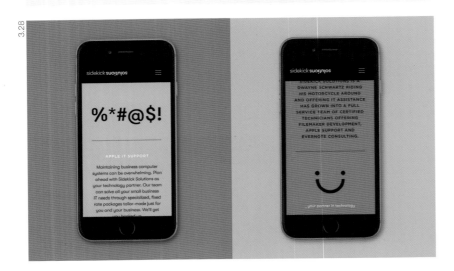

Centered copy is best used for title pages, invitations, and small amounts of copy. It really slows the eye down, so use these pauses to your advantage—make logical line breaks at the end of each line in a centered heading. Be aware that some development software for responsive websites may only allow centered text.

The ease of creating **centered alignment** often seduces new designers, but its classic symmetry leads many to overuse this format. The first time I summoned the courage to ask a professor for a portfolio review, I was told the biggest issue was that everything had centered text. It was so obvious but had never occurred to me. The embarrassment was enough for me to redesign all the projects into asymmetrically balanced layouts.

Contour formats follow a shape or image. This technique grabs attention. The trick is to keep the margins between the shape and text consistent. But it's difficult to read. Have you read the words on the Nike poster text yet? Maybe that's the point, it's a puzzle. Contour designs are appealingly playful. Still, imagine how slowly the eye must read the words in this delicious ice cream article.

Asymmetrical formats are used in designs to make dramatic statements. Each line, word, or letter is placed in a unique position. This format is sometimes used with elaborate typefaces to achieve better balance among letterforms. The reader needs to search for the beginning of each line, providing thoughtful pauses between the words. This theater poster is cleverly designed using a collage image and several type formats (from the top: contour, left, vertical, justified, left, and centered).

EDIBLE TASTINGS

IF you've ordered ice cream for dessert at a Boston-area restaurant in the past 20 years and loved it, chances are it came from Christina's Homemade Ice Cream in Cambridge. Using only cream, milk, and sugar as a base, owner Ray Ford creates custom flavors from ingredients stocked in his spice shop (adjacent to the ice cream store). Over 400 area restaurants buy from him weekly. In addition to the typical chocolate chip, coffee, and French vanilla, Christina's menu boasts flavors like green tea—made with real Japanese matcha powder—coconut cream, Gran Marnier, cinnamon rice pudding, and carrot cake. But in summertime he sets the intense flavors of dried spices aside in favor of more delicate, subtle, fresh ingredients. Beginning in early June, Ford orders herbs and flowers from Eva's Garden in South Dartmouth, Massachusetts, and begins making small batches of seasonal flavors like fresh rose petal, fresh chocolate peppermint, anise hyssop, honey lavender, and sage. Available in limited quantities and for just a few weeks in June and early July, these flavors sing of the season, so be sure to make a trip to Inman Square to seek them out. But in case you miss these fleeting flavors, here's a nod to a few of Christina's standards, just as delightful as the seasonal varieties, but available year-round.

Burnt Sugar
Ford "burns the sugar black" outside in foil pans set on top of a barbeque grill to reduce the brown sugar to a thick syrup, then mixes it with cream and milk. Its bittersweet amber flavor is much like the top of a crème brulee.

Khulfi
A mango, coconut, ginger-, cardamom-infused pistachio-studded ice cream. Unlike the traditional Indian dessert, which is thick, icy, and dense, Ford's version blends the flavorings of khulfi with the texture of traditional American-style ice cream for a smooth and creamy exotic treat.

Sweet Cream
Made from just milk, cream, and sugar, this is the most straightforward of all ice creams, even more so than vanilla. Richly textured, creamy, and not too sweet. Pair with ripe local berries or cut store fruit.

Ginger Molasses
Freshly grated ginger and dark, treacly molasses evoke the best gingersnap cookie, creamy and spicy and rich. A perennial favorite; any time of year.

BY SARAH BLACKBURN. PHOTOS BY MICHAEL PIAZZA

Christina's Ice Cream
1255 Cambridge Street, Cambridge
617.492.7021 christinasicecream.com

Type Tips

Follow these guidelines for producing text layouts that engage readers and enhance comprehension of the content.

Typeface
— Use a serif font for body text, sans serif for titles, or vice versa.

— Limit the number of fonts to two per design.

— Don't use novelty typefaces for body copy.

— Use italics within body text only for emphasis.

— Never use all capital letters in sentences: PEOPLE WILL THINK YOU ARE YELLING.

— Simplify.

Contrast
— Add visual interest to your page by contrasting titles with body copy.

— Achieve contrast with size, weight, font style, separation, or color differences.

— Produce the best legibility with high-contrasting black text on white paper or backgrounds.

— Black type on red paper has poor legibility because of low contrast.

— Simplify.

Layout
— Left align large amounts of text for legibility.

— Keep the space between words and letters consistent.

— Separate paragraphs with a line space or indents, but not both.

— Limit text to forty-five to seventy-five characters per line width.

— Sixty-six characters (letters and spaces) is considered ideal.

— Vertical line spacing is called leading (pronounce as led-ing).

— Use the auto-leading setting as a good standard.

— Place every element on the page in relation to the edge of another element.

— Zoom out and view your page as a whole.

— Make your document balance, from top to bottom, and from left to right.

— Leave plenty of white space around the edges.

— Simplify.

Adobe Illustrator Type, Drawing, and Advanced Type Tools

Chapter 2 introduced you to the Illustrator toolbox, Shape, Line, and Color tools. Let's now develop those and add advanced drawing and type skills. We'll focus on the following tools:

The Pen tool will initially be challenging, but it is fun and rewarding once you have had a bit of practice. It is used for drawing.

The white arrow is the Direct Selection tool. It is used to select a particular point on an object for making precise adjustments to drawings. Keep in mind, the black Selection tool selects an entire object with many points.

The Type tool generally works in the same way as most Type tools in word-processing software. However, when you put your cursor over the Type tool, then press and hold the mouse button, you'll see variations. The most useful is the Type on a Path tool. This allows you to place text along any shape or line, such as circles, waves, or a line illustration.

You will frequently rotate and resize objects, so these exercises will show you advanced methods for the Rotate and Scale tools.

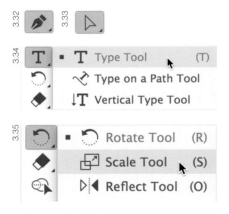

Design + Software Skills 3.1: Illustrator Typography Tools

Objectives
With this exercise, you will build software skills for using type and develop your ability to choose and combine fonts.

To Do
Set this quote three times, following the typographic rules from earlier in this chapter: "Life shrinks or expands in proportion to one's courage," Anaïs Nin.

Life shrinks
or expands
in proportion
to one's courage.
Anaïs Nin

1 **Draw a text box**
Start Illustrator, open a new document, and make it letter size. (To review a new document set up, see Chapter 2.1 Software Skills Step 1.)

Choose the Type tool from the toolbox.

Place your cursor on the file window, then click and drag to draw a rectangle with the Type tool. The rectangle automatically fills with Place Holder text.

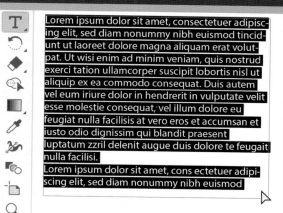

2 **Add your copy**
Type the quote into the text frame, using proper punctuation for this part of the exercise. Then select the text with your Type tool.

The Properties panel to the right will show the default type settings. If you do not see the Properties panel, go to Window > Workspace > Reset Essentials.

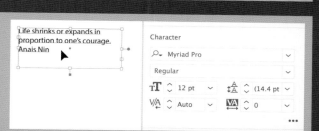

3 Format the copy

Use the Type tool to highlight the quote, then, in the Properties panel, make the font Avenir Light and the type size 12 pt. Leave the other boxes as default settings.

4 Italicize the name

Select "Anais Nin" with the Type tool, then change regular to Light Oblique in the font style on the Properties panel.

5 Change the text colors

Select the quote with the Type tool. On the Properties panel above the type options is a color Fill box, click on the box to bring up the color swatches. Select medium blue.

Now select her name with the Type tool. Click on the color Fill box and select the red-orange.

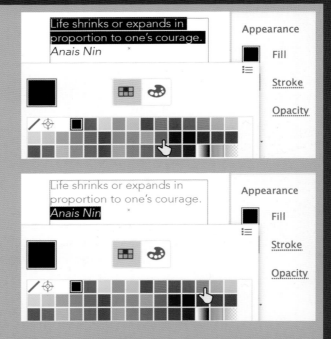

6 Add a special character

Refine your text by changing the *i* to an *í*.

Select the *i* with the Type tool, then choose Type > Glyphs.

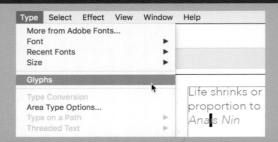

7 Make the Glyphs window larger to find the *i*. Double-click on *i* to change the symbol.

8 Click your cursor outside of the blue box to view your quote without the type frame.

Life shrinks
or expands
in proportion
to one's courage.
Anaïs Nin

9 **Design two more**
Try alternative type styles that enhance the message. Click on the text box with the black Selection tool. Press Command + C to copy and Command + V twice to paste two more text boxes. The second copy will paste directly over the first copy. Use the black Selection tool to move the duplicates away from each other on the page.

Select the type in the second box and set an alternative typeface, size, format, and color. To modify the proportions of the text box, use the black Selection tool. Click and hold a corner handle, then drag to adjust the size of the text box.

Life shrinks
or expands
in proportion
to one's courage.
Anaïs Nin

10 Modify the fonts

Highlight the text with the Type tool. Change the font, style, size, format, and color.

The version on the left uses Zapfino, 36 pts, centered format, for Life; and Futura, Light, 18 pts, centered format to the quote and name.

The version on the right uses Trajan Pro, Regular, 19 pts, right aligned. Now try your own font, size, and alignment choices.

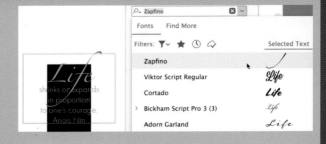

Tip

Did you notice the unusual placement of the period in the example at bottom right? This technique is called hanging punctuation and is used by skilled designers to give blocks of text very clean vertical margins. You can achieve this by applying Roman Hanging Punctuation to your selected and text.

First select your text and make it right aligned. Then, in the Paragraph portion of the Properties panel, click on the three dots for More Options. Then, click on the little hamburger menu in the top right and check the top setting for Roman Hanging Punctuation. Give it a try.

11 Print and Export

Well done. Now save your designs and export or print. Save, Export, and Print commands are on the File menu. I recommend you frequently press Command + S to save your work as you progress with a project. No need to wait until you complete a project to save it. And when finished, you may want to export to a pdf to share with friends or a client.

Export

Choose File > Save As then choose pdf from the Format pull-down options.

Print

File > Print or press Command + P to print.

12 Break the rules

On your own now—set this quote three more times, breaking the typographic rules to express the meaning of the words. These designs were created by students Brian Forte and Yuling Lu. We'll learn these techniques in Software Skills 3.2 through 3.5.

Tip

The coral-colored capital letters of the word LIFE in this example were adjusted with tracking and kerning. Tracking sets the space between capital letters throughout the word. Kerning adjusts the space between two specific capital letters to make spacing consistent.

To kern: Position your Type cursor between two letters. Then hold the Option key down and press the left or right arrows on the keyboard to increase or decrease the spacing. Professionals kern the letters in a logo or in an all-caps title, the goal is to make the space between uppercase letters consistent. But we never kern lowercase letters because as the space between letters increases, identifying each word in a sentence becomes more difficult.

LIFE shrinks or expands in proportion to one's courage. Anaïs Nin

"Life shrinks or expands in proportion to one's courage." ANAÏS NIN

Life shrinks or expands in proportion to one's courage • Anaïs Nin

Design Project 1: Lyrical Layouts

Set the lyrics of a song or a poem to demonstrate a sophisticated understanding of typography and layout. Use approximately fifteen to thirty lines of text, the artist's name, and the title.

Objectives

Demonstrate an understanding of classic typography guidelines.

To Do

1 Select a poem or lyrics.
2 Identify a typeface that reflects the nature of the words for the title and another typeface that has high legibility for the body of the text.
3 Set the text following the Type Tips and format guidelines in Chapter 3. These projects are by students Tao Xuzhi and Yuling Lu.

3.36

Moon River

Audrey Hepburn

Moon river
Wider than a mile
I'm crossing you in style some day
Oh, dream maker
You heart breaker
Wherever you're goin'
I'm goin' your way

Two drifters
Off to see the world
There's such a lot of world to see
We're after the same rainbow's end
Waitin' 'round the bend
My huckleberry friend
Moon river, and me

3.37

Comfortably Numb Pink Floyd

Hello?
Is there anybody in there?
Just nod if you can hear me
Is there anyone at home?

Come on now
I hear you're feeling down
Well I can ease your pain
And get you on your feet again

Relax
I'll need some information first
Just the basic facts
Can you show me where it hurts?

There is no pain, you are receding
A distant ship smoke on the horizon
You are only coming through in waves
Your lips move but I can't hear what you're saying

When I was a child I had a fever
My hands felt just like two balloons
Now I've got that feeling once again
I can't explain, you would not understand

I have become comfortably numb

I have become comfortably numb

Okay
Just a little pinprick
There'll be no more
But you may feel a little sick

Can you stand up?
I do believe it's working good
That'll keep you going through the show
Come on it's time to go

There is no pain you are receding
A distant ship's smoke on the horizon
You are the only coming through in waves
Your lips move but I can't hear what you're saying

When I was a child I caught a fleeting glimpse
Out of the corner of my eye
I turned to look but it was gone
I cannot put my finger on it now
The child is grown the dream is gone

I have become comfortably numb

Design + Software Skills 3.2:
Illustrator Shape and Type on a Path

Objectives
Learn to use advanced type and drawing tools.

To Do
Create a typographic target using the word "TARGET," a circle, and the Type on a Path tool.

1 **Draw a circle**

Open a new letter-size file. Select the Ellipse tool by clicking and holding your mouse over the Rectangle tool and selecting the Ellipse tool below. Once the Ellipse tool is selected, take your finger off the mouse.

Draw a perfect circle by holding the Shift key as you click and drag the Ellipse tool on the page. Make the circle about 4 inches (10 cm) across.

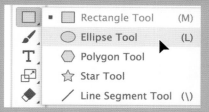

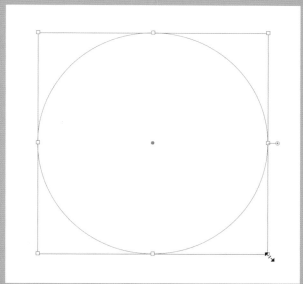

2   Type on the circle

Click and hold your mouse over the regular Type tool in the toolbox to select the Type on a Path tool below. Click the cursor at the top point in the circle.

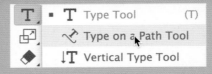

3   Format the text

Type "TARGET." All capital letters will give your type design a more uniform line.

Still using the Type tool, select the entire word. Then set the font to Impact, regular, and size to 18 pts. If Impact is not on your computer, download it from Adobe Fonts or choose another bold sans-serif font.

Change the color to red by clicking on the Fill square (above the type options in the Properties panel). Clicking on the color Fill square will bring up the color Swatches window. Click on the red swatch to change the text color.

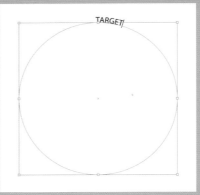

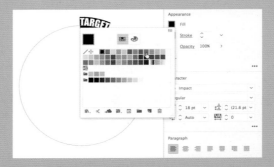

4 Copy, paste, and adjust

Using your keyboard shortcuts, press Command + C and Command + V to copy and paste the word "TARGET" around the circle. Notice that the last paste doesn't fit perfectly. Adjust this by selecting the entire circle of text with the Type tool.

Modify the font point size so that the entire word fits. Typically, the adjustment is only fractional, but this minor modification makes for a better overall design (your point size will probably differ from the number shown here).

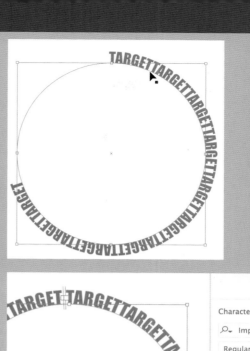

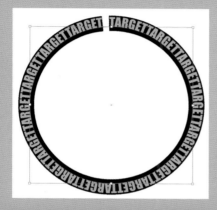

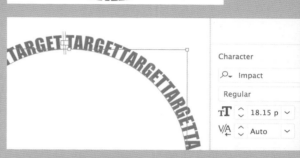

5 Copy a smaller circle

Choose the black Selection tool and click on the circle. Now place your cursor over the Rotate tool and choose the Scale tool from the pull-down options. Then double-click on the Scale tool to bring up the Scale window. Type "75%" in the Uniform Scale window and press the Copy button.

6 Build the target

Repeat the scale and copy process to create several circles. Double-click on the Scale tool, keep the percentage at 75%, and press Copy for each circle. The interior circle is a red circle you can draw and place in the center. To draw a circle from the center, choose the Ellipse tool, click at the center, and hold the Shift and Option keys while dragging to make a small circle.

Nice work.

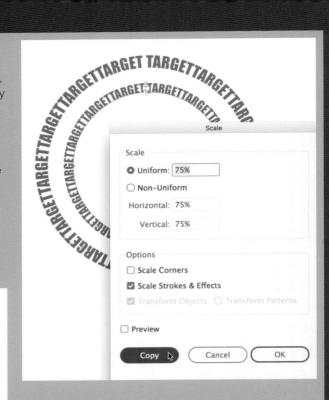

Design + Software Skills 3.3: Illustrator Pen and Fill a Shape with Type

Objectives

You will learn to draw with the Pen tool and fill shapes with text. You'll also learn creative uses for the Rotate tool.

To Do

Draw several triangles and rectangles using the Pen tool. Next, draw an arrow shape and use the Type tool to fill it with text. You will copy and rotate the arrow to create a new form.

1 Prepare to draw

Open a new file, letter size. Click on the default color squares, just above the Fill and Stroke colors in the toolbox. This puts the default colors, white and black, into the Fill and Stroke squares.

Click on the red diagonal-lined square, just below the large color squares, to change the fill color to no fill color.

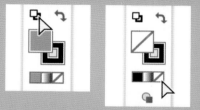

2 Draw a shape using the Pen tool

Select the Pen tool to draw a triangle using three clicks of the mouse. Click once, take your finger off the mouse, and move the cursor to draw a diagonal line. Click again, take your finger off the mouse, and move the cursor to create the second line of a triangle. Click on the first point to complete a triangular shape.

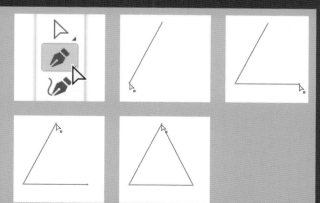

3 Practice

Draw several triangles. Try drawing rectangles and polygons. After you've filled the page with shapes, press Command + A to select all the shapes in the file, then press the Delete key to clear the workspace.

4 Draw an arrow

Draw an arrow shape with eight clicks of the mouse.

a. On the left of the page, click once, then move the mouse down about 2 inches (5 cm).

b. Click and release, then move the mouse to the right about 4 inches (10 cm).

c. Click and release, and move the mouse down a short distance.

d. Click and release, and move the mouse diagonally up, to make the arrow point.

e. Click and release, and move the mouse diagonally upward towards the left.

f. Click and release, and move the mouse down a small amount.

g. Click and release, and move the mouse to the left.

h. Click on the beginning point to complete the arrow shape.

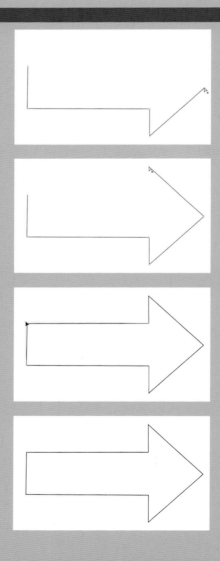

5 Adjustments

Once the entire shape is drawn, you can use the white Direct Selection tool to click on specific points and adjust the overall shape of the arrow.

Tip

If you have trouble selecting a particular point with the Direct Selection tool, it's because the entire object is already selected. Click off the shape, then click exactly where you expect to find a point on the line.

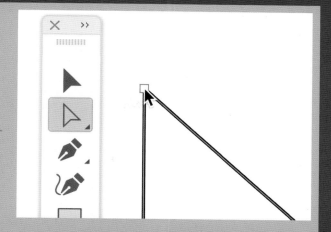

6 Fill with text

Select the Type tool. Click at the highest point on the shape. The arrow will fill with Place Holder text.

7 Format the text

Type the phrase "Who else is gonna bring you a broken arrow?" (Sweet lyrics from Robbie Robertson.)

Select the sentence and modify the font and size. This example is Museo Slab 500, 6 pts. Museo Slab provides a dense texture. If it's not on your computer, get it from Adobe Fonts or try another sturdy slab-serif font.

Select the formatted text with the Type tool, and copy and paste so the lyrics fill the entire arrow.

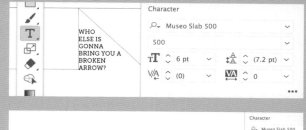

8 Change the alignment

While the Type tool is selected, press Command + A to select all the text. Select right-aligned paragraph format. Notice how this fills the shape more distinctly on the right. Look at the left-aligned format (above) for comparison. Reduce the font size to 4 pts to give the arrow a denser, darker appearance.

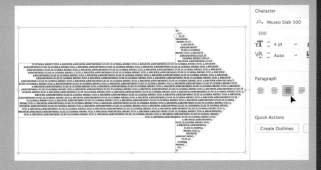

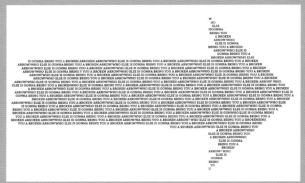

9 Resize

Reduce the size of the arrow. While the arrow is selected, double-click on the Scale tool, then type "50" into the Uniform Scale window. Confirm that the Option Scale Strokes & Effects is checked. Press OK.

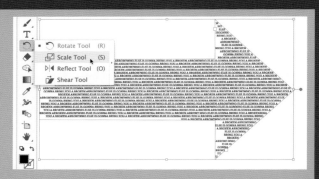

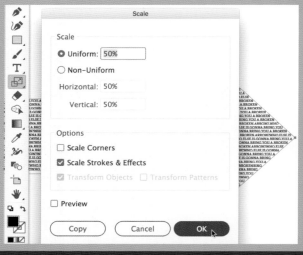

10 Rotate and copy

Make a mirror copy of the arrow: click on the arrow with the Selection tool, then double-click on the Rotate tool to bring up the rotate window. Type "180" into the Angle box, then press Copy (not the OK button). Using the black Selection tool, click on the new arrow and move it so that it faces the first arrow.

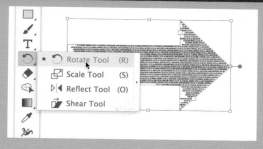

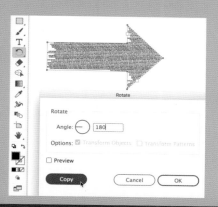

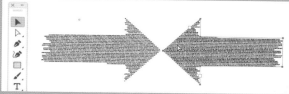

11 Rotate and copy

Hold the Shift key and select both arrows with the Selection tool. Double-click on the Rotate tool. Type "90" into the Angle box, then press Copy (not the OK button).

Move the two new arrows to construct a new shape.

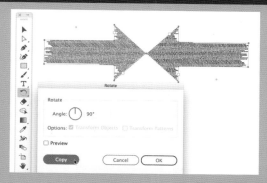

12 Finishing off

Hold the Shift key and click on all four arrows with the black Selection tool. Now click on the Rotate tool and move the cursor to freely rotate the entire shape.

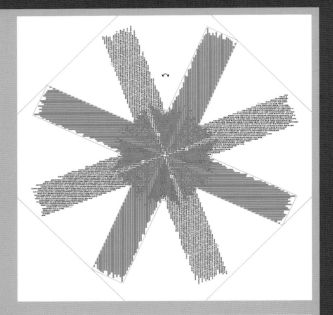

Design + Software Skills 3.4: Illustrator Pen Tool, Curves, and Type on Paths

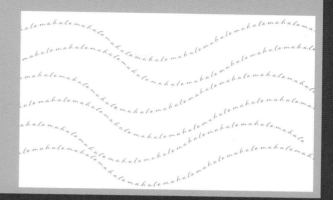

Objectives

You will learn to draw curved lines using the Pen tool.

To Do

Draw a wave using the Pen tool and use the Type on a Path tool to line it with "Mahalo," the Hawaiian word for thank you.

1

Draw a curved line

Open a new file, letter size. Set the Fill color to None and Stroke color to Black.

Select the Pen tool and draw a wavy line with four points. Click on the left of the page, then move the cursor about an inch (2.5 cm) to the right. Click and hold the mouse button as you drag to the lower right. A curved line will appear.

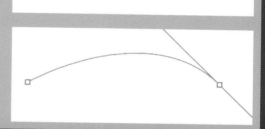

2

Click again

Click about an inch (2.5 cm) to the right and release the mouse button. This line's curve will mirror the previous curve.

Click again and hold the mouse button as you make another point, dragging to the right to create the curve.

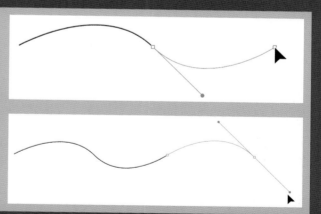

3 Click again

Click again another inch (2.5 cm) away.

You've drawn a wave. Don't expect perfection initially—you can undo (Command + Z) as many times as you like to practice your curved lines.

4 Adjustments

Professionals fine-tune by clicking on points in the line to make minor adjustments. Select the Direct Selection tool. Click on the line to make it active so that you can see the anchor points.

Click off, then click on one of the tiny blue squares and drag to adjust the position of the point in the line.

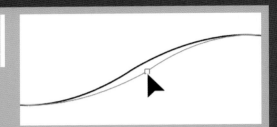

5 Adjustments

Notice the diagonal lines that run off the points on the line. These handles are used to adjust the depth of the curve. Click on the tiny blue square and drag right or left to make the curve shallower or deeper.

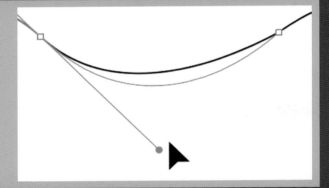

6 Add and format text

When you're satisfied with your wavy line, select the Type on a Path tool and click at the beginning of the line. Type "Mahalo." Choose a typeface, size, and color. This example uses Adorn Garland, 12 pts; make it blue. Adorn Garland has a casual elegance and lovely ascenders that create a varying height to the curve. If you don't have this font, you can get it from Adobe Fonts or use another script font.

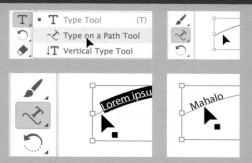

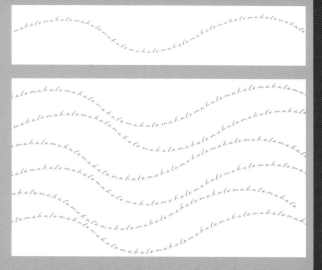

7 Finishing off

Highlight the text, then copy and paste (Command + C; Command + V) to complete the line with Mahalo text. You can copy and paste the line multiple times to make a design. These lines are copied and placed slightly offset. Aloha.

Design + Software Skills 3.5:
Illustrator Vertical Type Tool and
Convert Type to Shapes

Objectives
Convert type into shapes to expand their design possibilities as graphic elements.

To Do
Type a word using the Vertical Type tool. Convert the text to shapes using Create Outlines. Free rotate the letters and adjust their rotation.

1 **Add Text**
Choose the Vertical Type tool that is below the Type tool. Notice the cursor will appear sideways, and the letters will stack vertically as you type l-o-v-e.

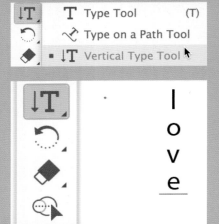

2 **Format the text**
Select the text and make the font Blackcurrant, Cameo, 60 pts. If you don't have Blackcurrant, you can get it from Adobe Fonts or use another bold novelty font. These letters are lowercase. If you are using another font, you may want to switch to all uppercase so the letters will be the same height. Select contrasting colors in the Properties panel.

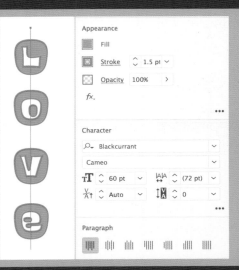

3 Create Outlines

Click on the text box with the black Selection tool. Choose Type > Create Outlines. After you've chosen this option, the letters cannot be edited as text. The letters are now shapes.

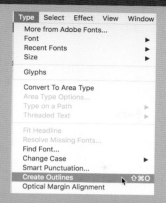

4 Ungroup the shapes

You can ungroup the letters using any of these three ways while multiple objects are selected:

a. Press the Ungroup button on the Properties panel or

b. Choose Object > Ungroup or

c. Press Command + Shift + G

5 Rotate the shapes

Click off the entire word. Select the Rotate tool, click on the L and drag your mouse to rotate the letter. Click on the other letters and slightly rotate them too.

Good work. Now using your name and a new font, practice converting type into shapes again.

Design Project 2: Slang Type

On your own now, use typeface and color to express the meaning of one of your favourite slang words. (Keep it clean!) These designs were done by students Jessica Hudson, Wasin Kittiwan, Stephanie Ploof and Pieter Melotte. Notice that no pictorial images or geometric shapes are needed for these expressive designs.

Design Project 3: Lyrical Layouts

Create a second layout using the lyrics or poem from Project 1. Experiment with type to express the meaning of the words. Using type forms as graphic elements has an inspiring history. Here is French poet Guillaume Apollinaire's 'Il Pleut' ('It's Raining') from 1916. The letters sprinkle like rain down the page.

For this project, use only type and other keyboard symbols, not images. Since you practiced the traditional type design guidelines in the first project, you may break a few rules here, if it helps to convey your feelings about the words. Use the type drawing techniques that you learned in exercises 3.2–3.5 to construct your design.

Students Raisa Acloque designed "Fade to Black," Ray Yu designed "Pumped Up Kicks," and Tao Xuzhi designed "Moonriver."

Major Points Summary

— The language of typography includes many useful terms that are illustrated in the diagram at the beginning of this chapter (see page 66).

— The term "typeface" refers to the overall character set of a particular design. It is used to describe the design of the letters.

— "Font" refers to the complete set of characters in any given size and style, and the software version of a typeface.

— The major categories of typefaces are serifs, sans serifs, scripts, and novelty.

— Serifs are the little notches at the ends of letters.

— Serif typefaces have been in use for 2,000 years. They are considered conservative in style and highly legible.

— Fonts designed by typographic groundbreakers, such as Garamond, Baskerville, and Bodoni, remain popular hundreds of years after they were created.

— Serif typeface categories include Old Style, Transitional, Modern, and Slab Serif.

— Designers often combine two typefaces on a page; avoid combining typefaces from the same category because they look too similar, and their use can appear to be a mistake.

— Leading contemporary typographers, such as Matthew Carter and the team at House Industries, often design with computer displays in mind, but they also use hand-drawn, traditional media in their designs.

— Mono-weight letters are more legible on digital displays.

— Novelty typefaces provide designs with immediate tone, but only use them for titles or small amounts of type because they can lack legibility.

— Type families provide variety for a typeface: roman or regular, italic, bold, and condensed are typical varieties. They can be combined effectively because they share characteristics such as x-heights.

— Justified type formats align on the left and right edges. This format has an orderly appearance and is highly legible.

— Left-aligned type format is also very easy to read. The edges align on the left and have a ragged right edge, minimizing hyphenation.

— Right-aligned type can be used for a few lines of type to balance a page. A complete paragraph or more would be unreadable, as the eye never knows where the next line begins.

— Centered type is a formal format. Take advantage of the pauses created by the line separations and make meaningful, or at least logical, line breaks.

— Asymmetrical type layouts really slow the eye down, so use them thoughtfully.

— Contour formats align to the edge of an image. Make the margin between the text and image consistent throughout.

— Type tips for fonts, contrast, and format are summarized (see page 78).

Software Skills Summary

— Illustrator type and drawing skills.

— Keyboard shortcuts.

— Skills: text frames, type options bar, color swatch window, glyphs, kerning, type on paths, type in shapes, vertical type, default colors, draw curves, convert type to outlines, ungroup images, hanging punctuation.

— Tools covered in depth: Pen, Direct Selection, Type, Path Type, Vertical Type, Rotate, Ellipse, Scale.

Recommended Readings

Once you've mastered these typography fundamentals, continue developing your knowledge by consulting other books on typography.

If you love fonts too, I highly recommend, *Just My Type* by Simon Garfield.

View a video of letterpress type setting in action: http://globeatmica.com/

To read more about the history of typography, check out Timothy Samara's new *Letterforms: Typeface Design from Past to Future*.

Blackcurrant font by Rian Hughes for Adobe.

Images

Milton Glaser, Bob Dylan poster, 1967

Throughout his seven-decade career, Milton Glaser explored new graphic techniques and influences. His wide-ranging work never revealed a particular visual style, but his designs are all conceptually rich and often witty. In this iconic poster, Dylan's hair was inspired by Art Nouveau arabesques, the flat colors by Japanese prints, and the black outlines in comic books. The black profile is a bit of a tease—the details must be filled in by the viewer's imagination. You can thank Milton Glaser for the I♥NY logo and its vast number of adaptations.

4.01

More than Meets the Eye

by Robert Bean

Recent Work by Scott Conarroe

Attracting an Audience

Scott Conarroe's photo of the Bixby Creek Bridge along the coast of Big Sur, California, evokes a variety of reactions, from thoughts of the sea's beauty, to the instability and dangers of the cliffs. The unusual placement and direction of the text intrigues us further.

What makes you stop, look, then read a graphic design? My bet—the image grabbed you. Most often, designs use photographs and illustrations to get our attention, and only then, after we've been hooked, do they provide information. A well-chosen image communicates on multiple dimensions: subject, mood, issue, humor, and/or information.

Great design relies on excellent images. Students have ready access to digital photographs, yet few guidelines for their selection and use. This chapter will give tips on finding the best image for the job. You'll learn the meaning of those baffling file formats, and you'll discover how to reproduce images at the best possible quality. Let's go beyond settling for the first image that matches our subject – let's get the ideal image for our design.

And, let's go beyond using photography exclusively. Illustrations can be produced in a variety of mediums: ink, pencil, collage, digital, paint, cut paper... The list goes on. The handmade quality of illustrations can be advantageous for many subjects.

Illustration

When are illustrations more beneficial to a design than photography? Illustrators can eliminate details. Due to minimal details, this book cover reflects the mysteries within the suspenseful novel *The Man Who Walked on the Moon*. Readers may then be surprised to learn the book is about years of hiking on a Nevada mountain called the Moon.

In the cookbook *Lateral Cooking*, these iconic pastries are easily identifiable using simplified (and charming) two-color illustrations.

On the other hand, illustration can be used to enhance details. Noroshi is a Japanese restaurant that serves regional noodle dishes from Hokkaido, which is famous for its local ingredients. When Noroshi opened another location in Tokyo, their detailed designs demonstrated the use of these indigenous flavours.

OfficeUS Manual is a large and extensive book that provides a guide to American architectural workplaces in the last 100 years. In these two spreads, the blue pages show furniture illustrations used to plan spaces. Would you guess the illustrations in the top right are from 1890, while the images in the lower spread are from 2014? This example demonstrates how illustrations can have a timeless quality compared to photographs, seen on the right.

Photographs, particularly when they include people, may look dated quickly because of changes in fashion. It's not just clothing—hairstyles, makeup, even moustaches—can date an image. If your product will have a long shelf life, consider using illustration. This Los Angeles street scene looks like Southern California, yet we cannot date the image without details in the cars.

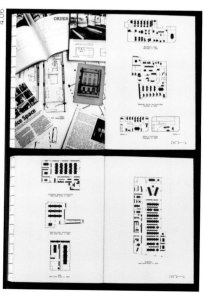

Illustration 105

Many subjects and processes cannot be photographed. Medical illustrators produce images that are essential to the health care industry—their work shows details of body parts, functions, and chemical processes. Keith Kasnot is a medical illustrator who works with software and traditional media to produce gorgeous images that make visual representations of things we can't actually see. As a certified medical illustrator, he is able to interpret complex biological systems for a general audience.

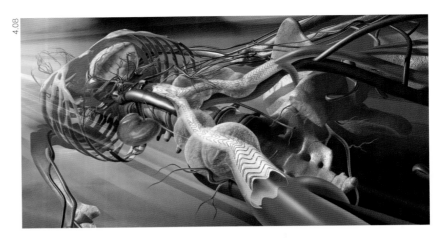

Now that photography is ubiquitous, the mere use of illustration attracts attention. You can use illustration to provide your designs with a fresh look. The poster for the play *The Whale* looks like a lithograph print, but it was drawn and modified in Adobe software. The bold illustration initially seems to be a fully dressed woman diving, but within a moment we are charmed by the images that emerge from the negative shapes that connect the image and the title. The text in the top-right corner reads, "Some tales swallow you whole." Who could resist?

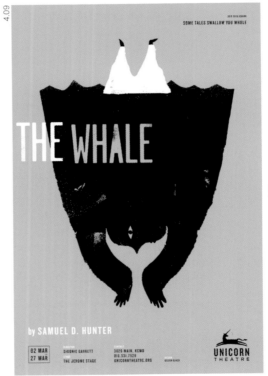

The handmade quality of images is often used to provide notes of authenticity, humor and character. The branding of the Battersea charity in England is notable for the use of watercolor and hand-drawn features in their logo. Here, we see a few of the ten cats and dogs that make up the logo system. This variety of faces is used to brand a wide range of services and represent the motto, "Here for every dog and cat." These faces are notable for a couple of reasons: they don't have features yet remain expressive and have a sense of individuality. They also avoid showing the animals as injured or ailing.

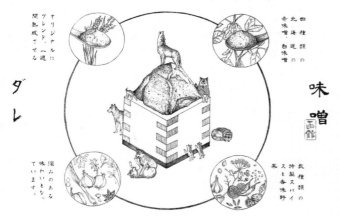

Fantasy and surrealistic images are made possible by the use of illustration. Styles vary with the themes, from photorealistic science fiction to gouache cartoons. This illustration for a ramen restaurant honours the extinct gray wolf from the Hokkaido region of Japan.

Notice how intertwining fingers create the number 15 for this anniversary poster. This not only adds visual interest—the integration of the copy with the image engages and informs the audience. This poster is proof of their tagline in the lower right: Always original.

The geometrical illustration on this early electronic music concert poster indicates that the music is retro yet also futuristic. Squares represent the basic form of electronic machines, and the multiple and different opacities mimic volume bars while creating a sense of variation and dynamics. The colors: conservative blue, creative purple, and an edgy bright blue enliven the abstract shapes.

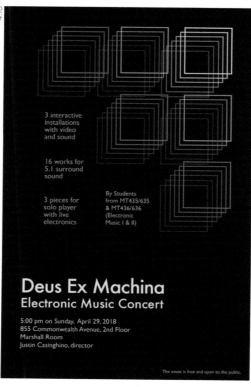

Illustration   107

Herbert Matter, Swiss tourism poster, 1935

Herbert Matter was a lucky young man. He studied painting in Paris under Fernand Léger and later designed posters with AM Cassandre. At age twenty-five he returned home and worked for the Swiss National Tourist Office, where he applied photomontage and Modernist techniques to poster design.

Matter pioneered the use of collages with dynamic scale changes and integration of type and image. Notice how this poster shows a sense of movement and evokes the experience of travel through Switzerland's mountainous roads. His photo collage of three mountains of various heights and multiple planes contrasts with the still, blue sky. The type is slanted, emphasizing the alps. This poster shows Matter's innovative integration of areas of color into black-and-white photography. His new approach explored reducing photographs to visual symbols, then combining images and words in surprising ways.

Photography

Have you ever found yourself examining an image, wondering whether it is an illustration or a photograph? Look closely at this coffee package. What initially appears to be a photo is an illustration of the macaw who roams the gardens at The Grounds in Sydney. The line between illustration and photography has blurred due to the capabilities and widespread use of digital photo-editing software.

Until recently there was a long-standing trust in photography. Prior to affordable desktop photo-editing software, most people believed photographs could not lie—photos were considered factual records. Police departments employed professional photographers who recorded crime-scene evidence as fact for use during trials. But now, digital technology has changed the reputation of photography as truth. Prior to this period, subscribers were outraged when a revered international magazine modified a photo of the Giza pyramids so that they appeared closer together—simply to fit the vertical cover format. That's messing with one of the greatest architectural feats on earth to suit the needs of a magazine cover! The magazine apologized, and its current policy prohibits manipulating elements in a photo to achieve compelling graphic effects. We can make the sky over a resort appear bluer. Is that an ethical choice when the locale is typically rainy? Be thoughtful with your choices and cognizant of the role (appeal?) of photography-as-truth when using this powerful software.

4.15

4.16

4.17

4.18

When gathering images for this book from across the globe, I encountered many examples of collages. These three, from Toronto, Barcelona, and Jacksonville, Florida, show range and experimentation. Why are collages becoming more common? Perhaps a cause is the number of photographs produced since the emergence of phone cameras. For the last decade most of us always carry cameras. It would be interesting to note the numbers of photos produced prior to 2008 and now. Notice also how these examples combine photography and illustration.

Image Resolutions

Have you ever sat at the computer waiting and waiting for an image to appear on a screen? The probable reason for the delay is the image file size is too large for fast transmission. As digital-image resolution increases, image quality improves, but file size grows too. As this happens, software processing time increases. This has an effect on our work, as we create designs, and on how well the image is displayed in the final product.

The typical resolution values you'll see are 72 dpi and 300 dpi. Dpi, an abbreviation of dots per inch, refers to the number of dots of ink on printed paper. When printing, more dots of ink per inch create sharper images: there are typically 300 dots of ink per inch. Resolution refers to the dpi number and indicates the sharpness of the photo. We call 72 dpi low-res (low resolution) and 300 dpi hi-res (high resolution).

These numbers, 72 and 300, also correspond to ppi, or pixels per inch. Pixels are tiny squares of light emitted from your computer screen. There are 72 pixels across each inch of a typical computer screen. This varies with high-resolution screens whose pixel rates can be double and triple this number.

Low-res images aren't always bad. Since most computers display images at a resolution of 72 ppi, these images look fine on computer screens and they load quickly. But the same image would look fuzzy when printed on a 300-dpi printer.

In general, digital images will be 72 dpi for screen displays and 300 dpi for printed publications. As you enter professional studios, these typical numbers may be modified depending on the media being produced. The zebra on the left is 300 dpi; the photo of the zebra on the right was saved at 72 dpi. Look closely to see where the image quality differs. Can you see the edges of pixels along the black stripes?

For consistently high-quality image reproduction, print designers typically work with 300 dpi images. In professional situations, some images will be scanned at higher resolutions, such as 1,200 or 2,400 dpi, for the final printing. However, the designer may create the layout using a 300-dpi image as a more reasonable file size for working in the software. When ready for press, a higher resolution version is linked to the file. You'll see that as our projects become more complex, the software will take longer to process the commands with large file sizes.

Tip

Because software usually links to images, we have best practices for working with image files. Get into the habit of starting each project with a folder, and place all your images and your software file into the same folder. This way, your images will be linked to the file, and will always display or print at their best quality.

TIFF, EPS, JPEG, GIF, PNG, and PDF File Formats

You've seen the terms TIFF, EPS, JPEG, GIF, PNG, and PDF, – but come on, does anyone really know what they mean? Yes, good designers know these terms and understand how to use them to create better designs. You can, too.

High Quality (printing)
TIFF (Tagged Image File Format)

TIFFs are used for high-quality images that are to be printed. TIFFs are large files and usually uncompressed. The term "uncompressed" means none of the original image data is eliminated to make the file smaller when saved in this format. Consequently, TIFFs maintain high image resolution. Most of the designs in this book were saved as 300 dpi TIFFs. Check out the zebra on the left.

EPS (Encapsulated Post Script)

Similar to TIFFs, EPS files are uncompressed, high-quality images. Currently, they are used less commonly than TIFFs, but when writing this book, I found them very convenient because they are editable in Illustrator and can produce smaller file sizes.

Compression Formats (screen display)
JPEG (Joint Photographic Experts Group)

JPEGs are used for photographs and complex illustrations intended for screen displays. This is the most common compression file format. File compression results in smaller file sizes, but it eliminates some details and colors. JPEG files can be saved with low, medium, and high quality. High-quality JPEG files will eliminate less information, but the file size will be bigger. When JPEGs are saved at 300 dpi and maximum quality, they can be printed, as are some of the images in this book. After you save an image as a JPEG, you can't restore the original quality, so keep a copy of the original file; the file extension is usually shortened to .jpg. The zebra on the right is a high-quality 72 dpi .jpg file.

GIF (Graphics Interchange Format)

This is a low-res, compressed file type for screen display only. GIFs are used for text, logos, and charts on digital designs, such as websites and apps. GIF compression eliminates variations of colors to reduce the file size. GIFs are great for graphics that have few colors. The menu graphics on my website are saved as GIFs. Don't save photographs as GIFs because the compression will reduce smooth gradients to bands of colors.

PNG (Portable Networks Graphics)

PNGs are a compression format that preserves transparency around a graphic, especially helpful when placing logos over a photograph.

PDF (Portable Document Format— the Glorious Hybrid)

PDF files accurately display all the characteristics of a design, even unique typefaces, without requiring the original software or fonts to be on the recipient's computer. They are used for sending designs via email, for large graphic and text documents that are available on websites, and for cost-effective printing. PDFs can be compressed with low, medium, or high quality for screen or print, and they are often created and viewed using Adobe Acrobat software.

Acquiring Digital Images

When you shoot your own digital images, you can open them on your computer with apps such as Photoshop and Adobe Bridge. You can then export images as JPEGs or TIFFs for use in Illustrator and InDesign. Additionally, designers can retrieve images from the internet, or you can scan printed material. Here are suggestions for getting the best-quality images from these sources.

Internet Images

We all use Google to search for images. For professional design purposes, the quality of images found using Google can be mixed. Designers use Google image search primarily for research purposes, but let's refine your search strategy to find better-quality images.

The first step in choosing high-quality images from Google is to look at the file-size information listed just below each thumbnail photo. When you mouse over an image, you'll see numbers below the image title. Let's focus on the second image in the top row: 2386 × 2983. This number is the file size in pixel dimensions. These numbers indicate very good image quality. A general guideline is if the dimensions are over 1,000 pixels, the image quality is good enough for research.

A 300 dpi file has 300 pixels per inch. So, divide the dimensions by 300 and you'll learn that this example's dimensions are approximately 8 inches × 10 inches (20.32 cm × 25.4 cm). Just by looking at the numbers we immediately know this image has good reproduction quality for print.

Many file sizes will be much smaller, especially when they're saved at 72 dpi, and therefore only suitable for screen display. The third image in the top row is 650 px × 365 px, approximately 5 inches × 9 inches (12.7 cm × 22.86 cm), which is much smaller than the previous example. And at 72 dots per inch, it is suitable only for screen display, not print. Google's advanced image

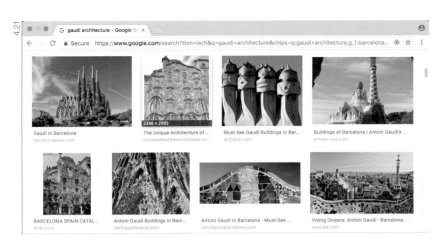

search allows you to eliminate small and/or medium file sizes from your searches to retrieve better-resolution images (but you will see fewer results).

You can download an image you'd like to use for research by clicking on it and dragging it to your desktop. Immediately check the image size and resolution to determine if it is good enough to use. In Photoshop, choose Image > Image Size (or press Command + Option + I). Confirm that the file size is 1 MB or more by looking at the resolution—is it lo-res (72) or hi-res (300)? Occasionally, you will see numbers in between 72 and 300. When the file size is above 1 MB,

it is very good quality. Print quality images are often 25 MB or more.

If the file size is well above 1 MB and the resolution is 72, you may have a hi-res image. To confirm, note the top number, which is the files size, and change the resolution to 300, then reduce the dimensions from 33.139 to 9.759. The other dimension will adjust proportionately when the link is active as you see here. The dimensions are adjusted to the second decimal position to get the file size back to the original number. Remember, these typically are copyrighted images, and students can use them for educational purposes only.

Image Banks

The most common resource for digital images used in graphic designs are internet image banks. Popular sites include www.GettyImages.com and www.masterfile.com. These websites contain searchable databases of everything from photographs and illustrations to videos and music. To find images from these sites, type a keyword into the search window, then sort through thumbnails of images that match your search criteria.

Regardless of image bank, all the images are copyrighted to protect artists' work from illegal use and ensure compensation. Most image prices are based on image size and resolution. **Rights-managed** image fees are based on the intended use—the price is calculated based on the commercial nature of the use, whether it will appear in print or online, its size and position, geographic range, and length of distribution. The cost to use a photo for a full-page ad in a national sports magazine is higher than to use the same image in a privately owned local restaurant menu.

Royalty-free images are not fee-free. There is a one-time cost for the use of the image in accepted ways, forever. The cost is cheaper than rights-managed images because the photographer or agency holds the copyright and the designer is paying for the right to use the image. There is a risk related to the lower price. You may see these images in other places because practically everyone else is interested in lower fees too. You can use a download of a low-res image from websites to comp a design. A fee must be paid (typically by the client) for any commercial or non-educational use beyond the classroom.

Screen Grabs

To capture an image file of your entire screen on a Mac, press Shift + Command + 3. To select a portion of the screen, press Shift + Command + 4 and drag the cursor around the area you wish to capture. Each selection will produce a "Picture #.png" file on your desktop. The highest-numbered file is the most recently captured image. All the software instruction images in this book are screen grabs.

Some Bad News

Brace yourself for some disappointing information about internet images. After you acquire a 72-dpi image from the internet, do not increase its size dimensionally. The image will look blurry if it is enlarged. I know—disappointing. If you acquire a digital image with 300 dpi, you can enlarge it up to 120 percent. Any larger and it will appear fuzzy. But there is good news for students who want hi-res images for their class projects: scanning.

Scanning

Scanning printed material expands your sources. Printed images (and objects) can be scanned to create high-resolution digital images. When a scanner is attached to a computer and the scanner driver is installed, open Photoshop and choose File > Import. The scanner will be listed in the pull-down menu. Select the scanner from the menu that appears (often it's a multipurpose printer).

Expect to see slightly different windows for each scanner, but if you understand the basic guidelines, you can choose the best settings. Determine whether your image is a photo or illustration, color or black and white, then choose these original format settings in the scan window. Even if you are scanning for screen display, scan your image at 300 dpi resolution, then save a copy at 72 dpi. Often, the same images are used on websites and for printed material. Copyright applies to these images too.

Copyright

What a perfect term—copyright. This law protects artists from having their work used without permission or compensation. As members of the communication industry, we want these talented professionals to collect income and contribute to their chosen profession—it's part of the creative cycle.

For educational purposes, image copyright restrictions are generally suspended for class projects. Any time your work is intended for commercial purposes or profit, you must acquire image rights and pay for copyrighted images. Don't worry: clients typically pay the costs for all images.

Additional online resources for high-resolution images include sites that encourage sharing for non-commercial use, such as Flickr.com and Unsplash.com. Users can upload, manage and share their photographs. Read the specific usage guidelines before downloading. Unsplash provides a resource for free high-resolution images. Downloads are free, and you will be asked, but not required, to credit the photographer.

Coincidentally, there is an Unsplash image in this book, it is used in a student's album cover design. Get into the habit of noting the artist and source of images; you'll soon become knowledgeable about the field. Once you're on the job, follow the attribution conventions of your employer or publication for artist notations.

Your Own Photos

You probably take a lot of photos. When you wish to use your own photography in a design, there are three important tips: (1) Save and use a high-resolution version. (2) Avoid cropping too closely with your camera's viewfinder. The photo may not have the same proportions as the final design, and it's often helpful to have large margins on all sides. (3) Generally, it's best to put the light source behind the camera.

Image Selection

How do you select from the thousands of images out there? Go beyond the first image you find that fits your subject, and repeat these words: quality, quality, quality. Then ask yourself – what do you want the image to communicate?
Does the content
— accurately convey the message?
— help identify the subject?
— inform?
— set the desired mood?
— make you laugh?
— shock?

SEVEN HILLS PASTA CO.

BY **RAYNA JHAVERI** / PHOTOS **ADAM DETOUR**

On a small plot in an Italian village just outside of Rome, Nonna Lina raised chickens, cows and pigs. She gardened with the time and tenderness that were the mark of her generation. And most Sundays she'd spend with her young grandson, Giulio, teaching him the ancient alchemy of flour and water. Together, they'd combine the two with care to create one of the cornerstones of Italian cuisine: pasta.

Sharing her embodied wisdom came naturally. The bonds with land and food were akin to the bonds with family and community: strong, deep and built to last. The experience struck her grandson profoundly, but little did young Giulio Caperchi know he'd eventually make this way of life into his way of making a living.

He never thought he'd get into the food business outright, he was seldom far from it in some form or another. After university, he took a six-month course with a Michelin-starred chef at A Tavola Con Lo Chef, a culinary institute in Rome. Summers were spent working in neighborhood trattorias, doing everything from waiting tables and cleaning dishes to helping out in the kitchen. He'd developed an interest in food policy, farming and global climate change, but always thought he'd approach it from an academic standpoint.

He moved to London to pursue a master's degree in political theory and sociology and ended up meeting Carol Sogigian—a like-minded food aficionado from Shrewsbury who was there for her master's in marketing. But Italy kept calling: The two moved back to Rome when Giulio landed an internship with the

Food and Agricultural Organization of the UN. His passion to protect farming and preserve tradition continued to grow.

When Italy's financial crisis hit, the couple moved to Boston, where Giulio ended up doing research with Frances Moore Lappé (the well-known food and democracy policy writer who wrote Diet for a Small Planet and exposed the inefficiencies of modern agriculture). Something in this experience finally tipped him from food academia into food action. In partnership with Carol—now his wife and business partner—the first iteration of Seven Hills Pasta Co. was born.

They started by making taralli, a traditional wine cracker. It fared very well when they tested it at farmers markets, but was extremely labor-intensive and expensive to make. In keeping with their adventurous, resilient spirits, the couple learned, evolved and pivoted to making pasta instead. Though the fresh pasta market was saturated, there were virtually no local artisans making dried pasta. It was a perfect niche.

"We're in a food renaissance," says Giulio. "People are interested in where food comes from. It's a great moment to have a local food business in this part of the country."

Their pasta recipe couldn't be simpler: Flour and water are the only two ingredients. It's made from scratch, without preservatives, in their kitchen in Melrose. A soft dough is formed and kneaded in machines, then extruded through custom bronze dies that give it the characteristic rough texture that sauces cling to happily. The final step involves a 24-hour

Is the image sharp? Disregard images that are unintentionally fuzzy, which suggests low file resolution. Does the image lack detail in the darker or brighter areas? This suggests poor camera exposure. Look for images that are appropriate for your message and that have very good image quality.

Photo Cropping

Once you decide to use an image, the next consideration is fitting it into the page format. Cropping eliminates areas around the edges of an image, which can enhance the

ability to communicate a message. Notice the three photos in this article. There are three levels of crops and each tells us something different about this fresh pasta product. On the left, we see the packages, so we can find it in a shop. The center is a close-cropped version showing the pasta shapes as a colorful pattern. And on the right, in front of a coordinating pale green wall, we see the proud makers. Cropping is used to modify the focal point and ensure which area attracts attention. Together, these three cropped photos tell a story.

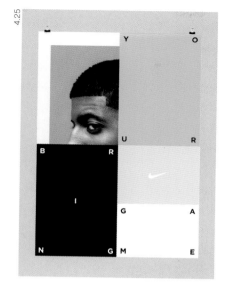

BOSTON STRONG

Aside from Boston's rich revolutionary history and (ahem) dozens of sports championships, it also plays a huge part in shaping the future. The city's innovative spirit is vibrant amongst cutting-edge design, robust research, and booming startup communities...not to mention at the 60+ universities. It's a cauldron of talent, new thinking, and opportunity, all operating at the crucial intersection of design, technology, and entrepreneurship. We're wicked proud to call Boston our home.

Basketball fans will recognize superstar Kyrie Irving even in this close crop. Other than the intensely cropped photo, this Bring Your Game poster is very restrained. Notice how the direction of his gaze leads us to the Nike logo.

Cropping is also used to fit an image into a particular space or page. Be careful! Don't carelessly eliminate details that help the image convey information. A good photo-cropping tip that applies to any occasion: try to keep body parts—fingers, feet, and ears—intact. Of course, there are exceptions to every rule, as we see in these appealing ads for an animal feed and supply company in Albuquerque, New Mexico.

Cropping landscape images can modify the sense of place, making them appear expansive. Alternatively, photos can be cropped to make the environment seem less grand and more intimate. This web page for a Boston studio uses three different crops to tell a more complete story of their location and hometown pride.

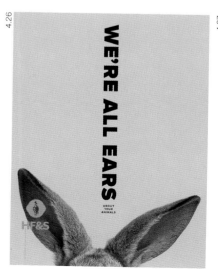

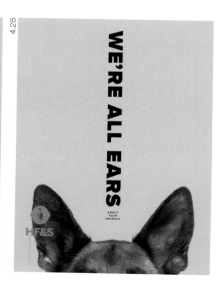

These two photos of wayfinding systems use wide crops to provide visual clues to locations and climates. On the left, the darker sky indicates that the new Riverside Precinct, on the east coast of Australia, encourages evening use and shows gathering spaces. On the right, a photo of the Waltzing Matilda Centre shows signage details, as well as the arid climate of the Australian outback.

These three photos for Taiyaki NYC have different levels of cropping to demonstrate their ice cream is "happy, fun, and delicious." First we see a cook making the fish-shaped cones (origin), then customers with the shop window featured (location), and finally the portrait of the cone (details) with a blurred but clearly happy customer.

Displaying Many Images at Once

Clients sometimes benefit from showing several photos in one design. While the form varies, similar strategies are used to maintain visual cohesiveness. In all these cases, the designs work because of a limited color palette, whether using black and white or a few colors. In the bicycle-themed grid, only grays, orange and yellow are used. Notice, also, the distance between objects is similar throughout each design. In some cases, there is a thoughtful variation in scale of the objects—see the big socks and small bikes, therefore each photo has a similar size on the page.

In the cheese board photo, twenty different foods are organized into a grid created with wooden boards. In the black-and-white photo collage, created for the history of an engineering firm, the contents span a century: some show people, others show architecture and building sites.

In this *We Can Work It Out* design for an article about women's gyms, the pattern is created with over sixty figures. The design works because only fifteen individual and similarly sized figures are used. Each figure is repeated, in different directions, using two to six variations. We will go into more layout strategies in Chapter 5.

4.35

4.36

4.37

4.38

Design + Software Skills 4.1:
Image Banks and Photoshop Good Crop, Bad Crop

Objectives

You will search an image bank for a photo and build Photoshop skills for modifying images. There are many image banks on the internet. In this exercise, we'll use Getty Images, one of the largest, with 200 million assets to choose from.

To Do

Find the best photo that fits one of the descriptions listed here. Crop the image so that the content is enhanced and try also to crop it so that the content is misleading. Why a bad crop? This is a quick and fun way to remind you to avoid cropping an image to simply fit a format in your layout. Cropping can strengthen or weaken the impact of an image in your design. Choose one of these topics

— Couples in love
— Couples fighting
— Time passing
— Animals interacting
— Athletes in motion

To acquire images in this exercise, use www.gettyimages.co.uk to search for photos. All internet image banks use very similar search techniques.

1 Search for photos

Go to www.gettyimages.co.uk. Type your search keywords into the window. Use keywords that narrow your search to find the best image for your topic.

2 Selecting

Browse through the thumbnails of matched images. If you don't see any you like, change your keyword selection. For example, you'll get more specific results from the phrase "skiing alpine" than "skiing."

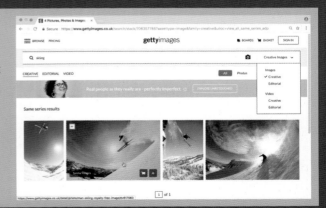

3 Previewing

Once you find a great image, click on it to enlarge your view. It will open in a preview window that provides prices, file sizes and rights information. Prices for royalty-free images depend on the size of the file. The cost is usually less for digital uses that will likely be smaller, lo-res files. For printing, the cost will be more expensive for the larger, high-resolution files.

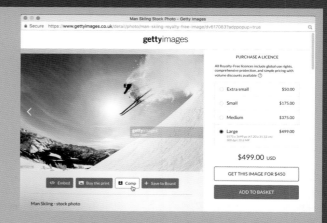

4 Pricing rights managed

Rights managed means the price of the image is determined by a calculation based on the photo's intended use. This includes the media in which it will appear, the client's industry, the geographical distribution, and duration of use.

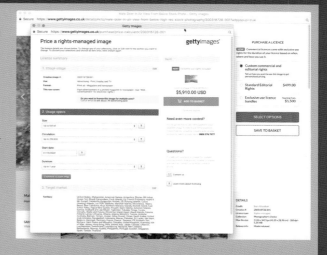

Tip

Image-bank photos are protected from illegal use with a logo watermark imposed on the image. When you work for an agency with a relationship with Getty, you may use a watermark-free image for comping, as you see here. You can use a download of a low-res image from the websites to mock-up a design. A fee must be paid (typically by the client) for any commercial or non-educational use.

5 Open the photo file

Download GettySkier.jpg from http://www.bloomsbury.com/graphic-design-essentials-9781350075047. Start Photoshop, select File > Open, click on the GettySkier.jpg file, and press Open.

6 Crop

Now we'll do a good crop/bad crop exercise. First try the bad one. Crop your image so that the visual message is impaired. Choose the Crop tool from the toolbox. Drag the cursor over the image. A box appears over your photo, and a gray outside border darkens the area that will be deleted. Adjust the selection by pulling the handles on the corners or edges of the image.

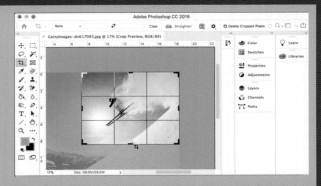

7 Crop

When satisfied, hit the Return key to apply the crop. You have permanently cropped the image to your selection. With this bad crop, the photo is less dramatic, and we've lost the charming details of the mountains in the distance.

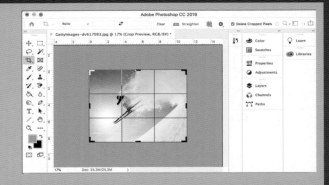

8 Revert

To go back to the last saved version of your image, choose File > Revert. Photoshop will bring back the original photo (or the last saved version).

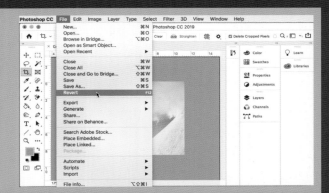

8   Crop

Now create a good crop to enhance the photo's message. You want a balanced image with the most compelling content to tell the story. Choose the Crop tool from the toolbox. Drag the cursor over the image. A box appears over your photo, and a gray outside border darkens the area you will delete. Adjust the selection by pulling the control boxes on each edge of the image. When satisfied, hit the Return key. The image has been permanently cropped to your selection. When final, press Command + S to save your edited image.

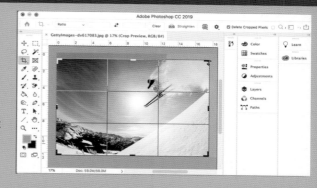

Tip

Rule of Thirds

A traditional photography composition strategy is called the "Rule of Thirds." This proposes that you envision three columns across, and three rows down a photo. Place major vertical features into the columns, horizontal features into the rows, and the focal point near an intersection of the lines. In this photo, notice how the mountains fill the bottom row on the grid. Expertly situated, the red boots, skis, and parka "pop" against the cool blue sky and are at the top-right intersection of the lines.

Design + Software Skills 4.2:
Photoshop Crop, Image Size and Resolution, Dodge and Burn

Objectives

Use the Rule of Thirds to crop this image. Learn how to use the Image Size window to evaluate image-size information and to resize a photo proportionately for high-resolution or low-resolution files. Learn to Dodge an image to make it lighter. Learn about the Burn tool that makes images darker.

To Do

Apply the Rule of Thirds to make a thoughtful crop. Change the size of a photograph proportionately. Save as a low-resolution file. Dodge and burn areas of a photograph.

1

Crop

Download the Giraffes.jpg file from http://www.bloomsbury.com/graphic-design-essentials-9781350075047. Open the image in Photoshop. Choose the crop tool and thoughtfully crop the image according to the previous tip, Rule of Thirds. Choose a good focal point. This crop aligns the grass in the bottom third of the image, the hillside trees in the top third, and the animals primarily in the middle row. Notice the galloping wildebeest is at an intersection of lines, establishing a fun focal point.

2 Image size

Open the Image Size window by choosing Image > Image Size (or press Command + Option + I).

It's wise to check this window every time you open a new image file in Photoshop. It provides essential information about the image.

The file size is at top center, and the dimensions are below. Set the units to your preference. Pixels are primarily used with digital projects; inches and centimeters are often used with print work. Note the resolution of this file is 300 dpi.

The dimensions are large; let's change them to a manageable 5 inches × 8 inches (12.7 cm × 20.32 cm). When the Link icon is selected, as you see here, you need only change one dimension, and the software proportionately changes the other dimension. Notice the file size, listed at the top, is much smaller, but the resolution remains the same.

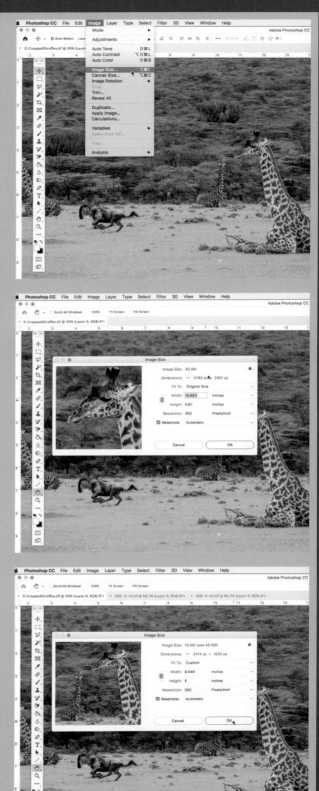

3 Make a low-resolution file

In the resolution box, change 300 dpi to 72 dpi. This makes the image size even smaller by eliminating data that provides fine details in the high-resolution version. Click OK. Your image is now resized proportionately.

Choose File > Save As and save it as a jpg file with a new name.

Tip

When you have a high-resolution original file, and plan to use only low-resolution files, always save the original for the length of the project. Once you save a file as 72 dpi, the only way to subsequently use a high-resolution version is to open the original 300 dpi file.

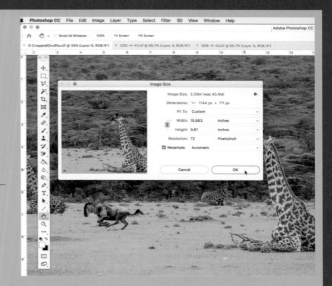

4 Dodge an area

Dodging lightens an area of an image. This tool can be used to show more detail in a particular area of a photo.

Select the Dodge tool from the toolbar. Notice the options at the top of the window. Usually, the default settings work well, especially 50 percent exposure. The goal is to achieve natural, unnoticeable adjustments.

The Options bar provides a Brush size window when you need to make the brush diameter larger. This brush size is 65 pixels. That's useful in this situation because it is a small brush size, which is necessary to dodge along the giraffe's head.

Select the Dodge tool, click at the nostrils and drag to the brow bone area.

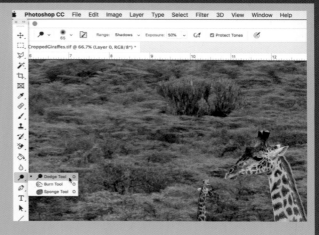

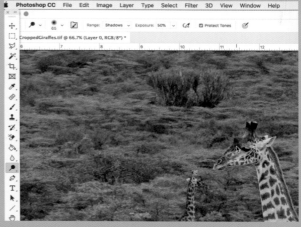

5

History

Can you see more details in the top photo, compared to the photo on the bottom (the original)?

Use the History window to compare both versions. Choose Window > History. The bottom row of the History window will show the Dodge tool. When that row is active, you see the edited version. When you click to the row above, you see the unedited image.

Evaluate the results. They should be subtle. If there are more details apparent on the head, and the edit looks natural, click on the dodge row to confirm the changes.

If not successful, click on the row above the Dodge tool to undo the edit. Give it another try. Using the History window to back up a step, or a few steps, is common in Photoshop. Backing up and adjusting your edit leads to better results.

6 Burn tool

The Burn tool makes part of an image darker. When text appears above a photograph, designers use the Burn tool to make the area behind the letters subtly darker. The words over the photo are then easier to read because the image is generally darker with less value change. Imagine this photo as a web page header. Let's burn the top third of the image (not the giraffes!), so the web page title will be more legible.

Select the Burn tool under the Dodge tool. Look at the Options bar above. Change the brush size to 300. A larger diameter covers more area and looks more natural. Click the Burn tool on the left of the tree area and drag the brush to the right. The hillside should appear subtly darker. Use the History window to compare versions. Evaluate the results—the original is shown in the top photo, and the burned photo is below.

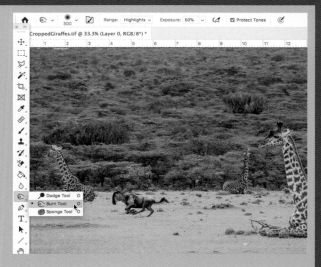

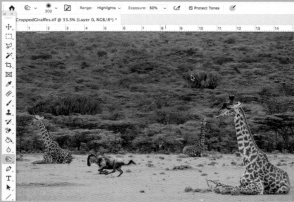

Tip

The Dodge and Burn tools use default colors to make the adjustments. You can customize these using colors from your photo for even more realistic results. Select the Eyedropper tool and click on a very dark area of the photo. Use the bent arrow above the colors to put the dark color in the lower square. Now click on the upper (foreground) square. Click the Eyedropper on a very light area of the photo. Now the tools will use those colors as the basis for darkening or brightening.

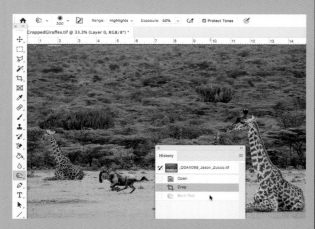

Design + Software Skills 4.3:
Photoshop Image Adjustments, Selection Tools, Hue/Saturation

Objectives
Learn the pros and cons of selection tools, adjust image exposure manually, and change hues in a photograph.

To Do
Edit the exposure, brightness, contrast and hues of an image. Select areas on an image with three different tools.

1 Acquire the photo
Download Bo-KaapHouses.jpg file from http://www.bloomsbury.com/graphic-design-essentials-9781350075047 to your computer. Open the file in Photoshop.

2 Image adjustments
The Image > Adjustments menu in Photoshop is loaded with great capabilities. For now, let's look at a common need: exposure adjustments. In Chapter 1, we introduced my favorite adjustment tool, Curves, an information rich interface. Conversely, the Brightness/Contrast window is a simplified interface. With no image information in the window, you simply trust your eyes to judge exposure. How do you know if your image needs an adjustment? Look for details. Adjustment is needed when either the darkest or lightest areas of the photo lack details.

Select Image > Adjustments > Brightness/Contrast from the menu. Check Preview so you can see the changes in the photo. Adjust the sliders to improve the image. If you're uncertain, click the Auto button. Note the values, press Cancel, and try to improve on the auto suggestions. Photoshop is very good, but I'll bet your eye can do even better.

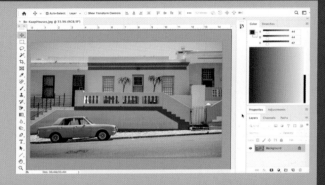

3 Quick selection

Let's explore three Selection tools to see their pros and cons. Quick Selection is seductive—it's a great name, and it's readily available on top of another selection. Let's try it.

Choose the Quick Selection tool. Adjust the brush size because this is a hi-res image. In the Options bar, change the brush size to 100.

Now click on the yellow house. Try to select just the yellow areas. Once you select one area, hold the Shift key down to select a second and third area.

What happens when you click on the yellow wall? It also selects non-yellow areas, such as areas with white paint and the window, right? Whenever you select something you don't want to use, press Command + D to deselect.

Press Command + D. This tool is too imprecise to work on this image. Let's try better tools for selecting.

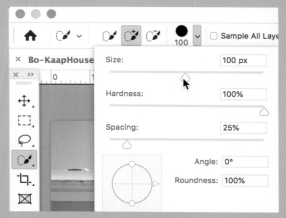

4

Magic Wand

Hold your cursor over the Quick Selection tool and choose the Magic Wand below. The Magic Wand selects similar, adjacent colors. Adjust the range in the Options bar's Tolerance window. Taking the time to adjust this number will pay off with more specific selections. Make the Tolerance 38. We want to only select the yellow areas of the house. The initial selection is promising because it does not include the window. Hold the Shift key down to select in the corner of this first area. Great. Now keep holding the Shift key down as you select the other yellow areas. Good selection. Notice how the white areas are not included.

5 Polygonal Lasso

If we want to select just the yellow house, the most precise approach is to use the Polygonal Lasso tool. Below the Lasso tool, select Polygonal Lasso. This is like the Lasso tool, but only selects straight lines—great for most buildings.

Click on the edge of the yellow house, take your finger off the mouse button, move the cursor down, click, move to the right, click, and so on around the edge of the house. Keep moving and clicking at every angle until you reach the starting point. Click at the beginning point to complete the selection.

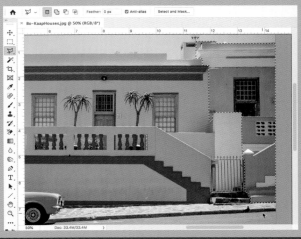

6 Copy

Let's copy this selection to another layer to make edits safely away from the original. Press Command + C to copy, then press Command + V. Notice copy and paste make a new layer.

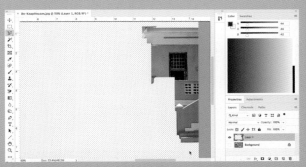

7 Change colors

With apologies to Cape Town, let's change the color of this house.

Click on Layer 1. Choose Image > Adjustments > Hue/Saturation. In the Hue/Saturation window, move the Hue slider to change the color of the house. This adjustment made the Hue value +26. Press OK. Now click the eye icon on the Background layer to see the three houses together.

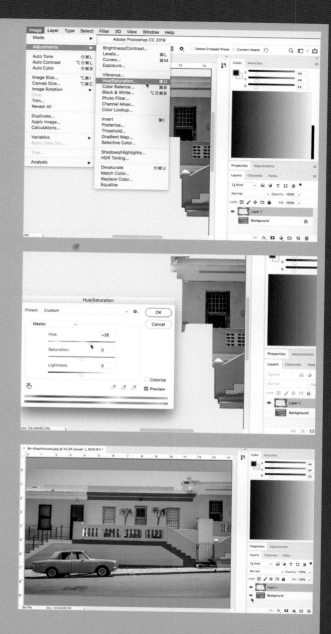

8

Finishing up

I don't think that new green is an improvement, so I toggled off the eye icon for Layer 1, clicked on the original layer, and went back to the Brightness/Contrast settings to adjust the exposure. I'm pleased with the adjustment; trying new approaches is valuable for designers. I wonder how your adjustments look.

Many thanks to Getty Images UK (skier) and Jason Zucco (giraffes and houses) for these beautiful images that brought us to other parts of the globe as we learned Photoshop.

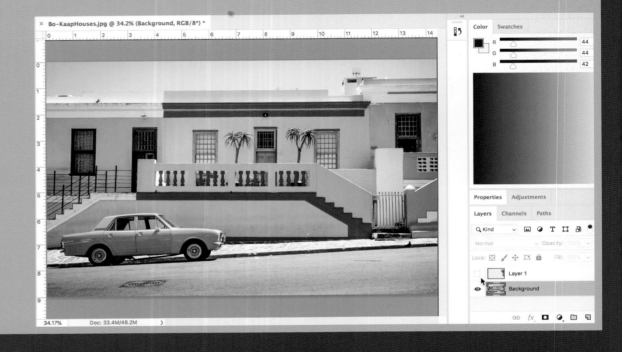

Major Points Summary

— Great design often relies on excellent images.

— The use of illustration rather than photography can be advantageous when used to eliminate details, enhance details, provide a timeless quality, or depict things that can't be photographed.

— The handmade quality of illustrations can provide humor, character, or authenticity.

— Photographs have evolved from being considered factual records of moments frozen in time to representations of genuine or idealized subjects.

— Digital images for screen display are typically 72 dpi, and printed images are typically 300 dpi.

— TIFF, EPS, and PDF are high-quality digital-image file formats used for printing.

— JPEG, GIF, PNG, and PDF are compressed digital file formats, and therefore they have lower resolutions, which are appropriate for screen display.

— Google Image search returns can be quickly evaluated to find the best-quality images.

— Image banks are databases that provide professional, copyrighted photography, illustration, clip art, sound, and video.

— Rights-managed images are priced according to the intended use.

— Royalty-free images are purchased once for unlimited use.

— The "Rule of Thirds" is a traditional photography composition strategy. Draw three columns and rows on the image. Place major horizontals and verticals into the columns and the focal point near an intersection of the lines.

Software Skills Summary

— Digital-image file formats and terminology are defined; scanning overview.

— Image bank search, acquire, price images, copyrights, rights terms.

— Photoshop advanced skills: crop, revert, resize proportionately, image-size window, set low-resolution, History, Dodge, Burn, Quick Selection, multiple selections, Deselect, Magic Wand, Polygonal Lasso, Hue/Saturation, Brightness/Contrast, Curves.

Recommended Readings

The *Communication Arts Illustration Annual* and *Photography Annual* are available in print and online at www.commarts.com. Read these to become familiar with leading and emerging illustrators and to stay on top of photography trends and resources. Illustrators and photographers that you enjoy will often be on Instagram, including those featured in this chapter.

This is a podcast by and about professional illustrators: https://www.svslearn. com/3pointperspective

Layouts

Saul Bass, *The Man with the Golden Arm*, 1955

5.01

The movie *The Man with the Golden Arm* marked the first time a comprehensive design program unified print and media graphics for a film. Saul Bass designed the logo, theater posters, print advertisements, and a groundbreaking movie title animation.

A single white vertical bar on a black screen begins the title animation. Next, three more bars join in, jaggedly moving to a staccato jazz soundtrack. Names of the actors interact with the jazzy bars, which then rotate horizontally to interact with the title, and rotate again for the production names. The music gets louder and more intense as the lines run vertically, converge and then form the jagged arm you see in the poster. With this film, Bass pioneered a new approach to film title design, where forms appear, disintegrate, reappear and transform through space and time. This Otto Preminger movie is about a heroin addict. As with all his designs, Bass was able to reduce his message to a single dominant image. He used freely drawn or cut paper, so the images appear casual, yet powerful.

Path Layouts

A good path layout may look like an effortless, spontaneous solution. But it's not. In a path layout, the designer skillfully arranges objects on the page so the audience's eyes are brought through all the contents in a meaningful sequence. This is a particularly effective approach for single-page designs, such as home screens, advertising, and posters.

A good path starts with a focal point. This is the area of the page where the eyes go first: a focal point attracts attention and encourages the viewer to look further. Let's face it, people are inundated with visual information. How do we get noticed without using nudity in every design?

Taiyaki is a Japanese ice cream shop that has expanded to New York, Boston, and London. Where do your eyes go first on this design? The unlikely fish-shaped cone, right? It is the largest, brightest area on an otherwise, white or nearly white, design. Then our eyes quickly move through the content. We notice the creamy ice cream and the hand leads us to the right, where we see a large logo that is read first. We've been engaged by the image and the logo, so we are inclined to continue reading the next line of text. Due to a successful path layout, our eyes are led throughout the entire design. Paths begin with an effective focal point and have good visual hierarchy of information. This design strategy ensures that the client's message is quickly and thoroughly communicated.

Capturing the Audience and Enhancing Communication

Where do your eyes go first in this cover design? Where do they go next? An effective path layout leads our eyes through the entire design (even when we don't know the language). The path our eyes take through the elements of this Italian magazine cover starts at the giant D—the biggest or brightest object on the page is usually the focal point. Then the illustration subtly guides our eyes through the layout with the positions, size and colors of the text. In this chapter we'll learn the art of layout: how to engage an audience and lead them comfortably through the essentials.

Effective layouts grab the audience's attention and hold it until they've read through the entire design. Layouts are tools designers use to communicate a lot of information cohesively, and they aid the enjoyment and comprehension of publications. We'll explore two different layout strategies that create effective graphic design: the path and the grid.

Focal Points

The focal point establishes an emphasis in the design that becomes an appealing entry to the path layout. To establish a focal point, you can use contrast to do the following:

— Make an object a bright color (as you see in the preceding example)
— Make the object dark and everything around it light
— Emphasize a compelling image, word or letter
— Make one object sharp and all other elements softly focused
— Give the object a different texture or gloss
— Make the object color and the rest black and white
— Change the value of an object to create contrast with the background
— Place an object in an unusual direction or position on the page
— Isolate the object on the page

This last tip is interesting: areas of white space have worked wonders. In this Citi scaffold-wrap, nearly half the design is empty. A large blank area in any graphic design is so unusual that it tweaks our curiosity and becomes an effective focal point. The tag line, "Live richly," is presented simply, to imply authenticity.

Bassett coffee bags effectively highlight their name by contrasting strong, slab-serif black text on a simple white package. The geometric pattern on the sides enlivens the package. White space, also called negative space, can work effectively with colors, or with subtle patterns. The expansive blank black space draws our eye, spotlights the word "Community" with large, contrasting white letters, and effectively guides us to the small rectangle of text.

Here is another clever use of negative space. The top of the image is over-exposed, creating a very light background that semi-conceals the Highline's H logo. Again, so much bright white space draws us into a design. Once this focal point attracts our attention, our eyes move along a path. We next notice the central figure, then the large 2, then the next largest text, Art, and so on through the important content. A successful path layout communicates information in a structured process.

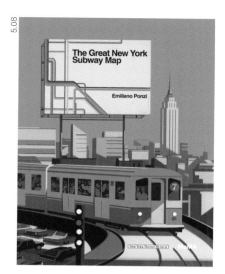

Integrating type and image

Place your main text near the focal point, or along the path, to integrate the text with the image. There are many ways to establish a good focal point, but remember it's part of a whole. If it's too strong, the reader will stop there and miss the rest of the design's content. On the other hand, if you don't establish a focal point, you'll end up with a confusing design. The audience won't know where to focus their attention and will move on. The goal is to begin the eyes on a path and then lead them through all elements of the design. Accents, such as supporting images, text, and/or logos, can be placed to guide the eye through the page. Repetition of colors, shapes, and textures also creates flow.

In this book cover about the New York City subway map, the title is placed into a billboard shape often seen from the Number 7 train on elevated tracks. Notice how the author and illustrator's name aligns to the horizonal lines. The designs in this book were shown in an exhibition, and the logos for the two museums are smaller and at the

bottom right of the cover. These are the last text to be noticed, but not missed. Effective visual hierarchy of information makes this a more comfortable reading experience.

Visual Hierarchy

Visual hierarchy refers to the relative importance of the elements on the page. It's essential to make these choices for your designs. The goal is to create a comfortable reading experience. Readers will subconsciously understand via your design choices where to look first, second, third, and you will lead their eyes through the important content of the design. In this advertisement, notice the variation in sizes and colors of the text. The headline is largest and brightest blue, while the secondary copy is also all capitals but slightly smaller and black. The supporting copy is upper and lower case, regular weight text. The tag line about shipping is, appropriately, the last item noticed. Four levels of visual hierarchy indicate the relative importance of the content.

Where do your eyes go first on this advertisement? Our eyes first see the focal point—the fisherman and fish. What path do your eyes

take through the rest of the design? Once we see the fish, we notice the logo in the top left. Our eyes then go to Sustaining, pulled by the larger size and higher contrast of the blue text. Next, our eyes move through the text and then to the supporting elements in the photo—the boat at the dock. Make thoughtful visual hierarchy decisions while designing with text. Establish an effective focal point, integrate the type with the image, establish visual hierarchy of information, and you will achieve an engaging and easy-to-comprehend design.

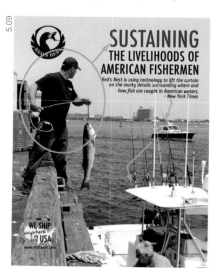

Design Analysis: Path Layout

Objective
Analyze the effectiveness of path layouts, recognize focal points and visual hierarchy.

To Do
Sketch the focal point and path of an existing advertisement.

Tip
Analyze the visual hierarchy and describe how the elements are arranged. Is this arrangement based on the elements' relative importance to the message? Find a well-designed advertisement with a path layout. Print or tear it out, then draw the path directly on the printed version.

Introduction to Adobe InDesign

Adobe InDesign is primarily used for multiple-page layouts. Oh, but it is so much more. InDesign is excellent for organizing images and for preparing files for professional printers (prepress). It has many of the capabilities that you'll use in Illustrator. Yet the best thing about InDesign is that it was developed to support graphic designers' work processes. Nearly everything we want to do in multiple-page layouts has been thoughtfully included in the software. You will use Photoshop to optimize your photographs and Illustrator to create logos, drawings, and single-page designs. But InDesign is essential software for multiple-page, text-heavy publications, and preproduction organization.

The InDesign Toolbox

You'll notice the toolbox looks like Illustrator's. The tools are similar, yet their capabilities are specialized for much more text and linked-image needs.

One new tool you'll use is essential: Frames. Frames provide structure for the placement of images and text. Yes, you can simply place images and text into a document, but when position is key, frames are important, and when document layouts are used repeatedly, frames are fabulous—copy and photos can be changed in the frames, and the layout stays the same. When you press and hold on the Rectangle Frame tool, you'll see it also provides elliptical and polygonal options. Notice the Frame tool looks like the Rectangle tool with an X inside.

Another thoughtful addition to the InDesign toolbox is just below the Fill and Stroke colors. Notice the tiny box and the tiny T. Click on one or the other to apply new colors to either the frame shape or the text.

An important distinction arises with Frames and Selection tools in InDesign. Use the black Selection tool to adjust the frame (size, position, etc.). Use the white Direct Selection tool to adjust the photo (size, crop, etc.) inside the frame.

We work in the Essentials Workspace. Set it initially under Window > Workspace > Essentials.

If at some point you can't find a window or panel, simply reset the workspace by choosing Window > Workspace > Reset Essentials.

Always start each new InDesign project with a folder and put all your image and .indd files inside. This is necessary because InDesign links your placed images to the .indd file. By placing your images into the same folder, the files will be linked and reproduce successfully.

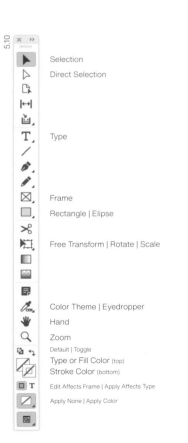

Selection
Direct Selection

Type

Frame
Rectangle | Elipse

Free Transform | Rotate | Scale

Color Theme | Eyedropper
Hand
Zoom
Default | Toggle
Type or Fill Color (top)
Stroke Color (bottom)
Edit Affects Frame | Apply Affects Type
Apply None | Apply Color

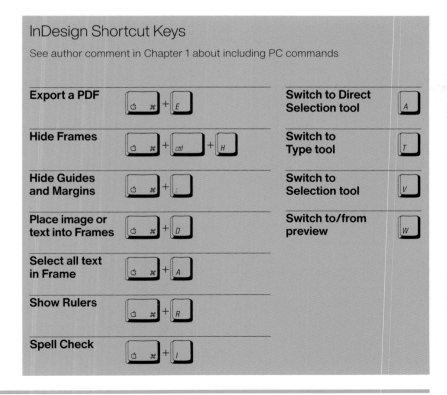

InDesign Shortcut Keys

See author comment in Chapter 1 about including PC commands

Export a PDF	⌘ + E	Switch to Direct Selection tool	A
Hide Frames	⌘ + ctrl + H	Switch to Type tool	T
Hide Guides and Margins	⌘ + ;	Switch to Selection tool	V
Place image or text into Frames	⌘ + D	Switch to/from preview	W
Select all text in Frame	⌘ + A		
Show Rulers	⌘ + R		
Spell Check	⌘ + I		

Design + Software Skills 5.1:
InDesign Place Image and Path Layout

Objective
Introduction to InDesign. Use Frames, place images, and format copy to create a path layout advertisement.

To Do
Place a photograph into InDesign and add text to create an effective path layout.

Tip

This photo treatment is called full-bleed. It means that the image goes to all four edges of the page. Bleed refers to the ink that extends beyond the borders of the design, and the edges must be trimmed after printing to produce these pages. Designs that are not full-bleed are often cost-savers because they can be produced without trimming the edges, reducing paper and production expenses.

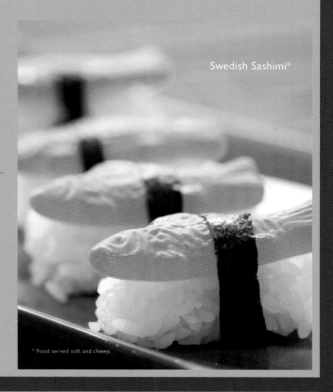

Swedish Sashimi*

* Food served soft and chewy.

1 **Make a new folder**
Create a new folder on the desktop for this project. Click on the desktop to make it active; choose File > New Folder. Click on the untitled folder to highlight it and type "SwedishSashimi." Always put your images and InDesign files in a folder, so the links remain constant.

2

Download the Image
Download the "SwedishSashimi.jpg" file from http://www.bloomsbury.com/graphic-design-essentials-9781350075047 and drag it into your project folder.

3

Create a new document
Open a new document. Choose File > New > Document.

In the New Document window, click off Facing Pages (this setting is useful for double-page spread projects such as magazines). Make it letter size, inches, one column. Press Create.

4

Draw a frame
Select the Rectangle Frame tool. You can draw frames freehand, but to be exact, just click once in the file window. Type 8 inches (20.32 cm) and 10 inches (25.40 cm) in the Frame window width and height boxes. These are common advertisement dimensions and allow you to trim off a white margin for the full-bleed presentation.

Choose the black Selection tool to click and move the frame into the center of the page.

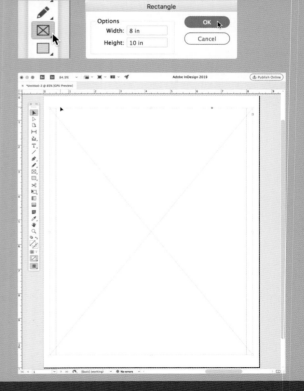

5 Place the image

Choose File > Place and select the "SwedishSashimi" folder, then "SwedishSashimi.jpg" from the file listing. Press Open.

Tip

As you get used to InDesign, use the keyboard shortcut to place images and text: Command + D.

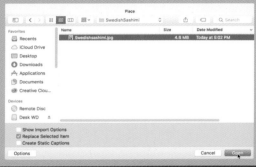

6 Move the image

You'll see the photo appear in the frame. Now choose the white Direct Selection tool to click on the image to adjust the position of the photo in the frame. Notice the box and handles are now red. This indicates you are adjusting the image, not the frame. Nudge the photo over (using keyboard arrow keys if you like) so that the sashimi are well situated in the frame.

7 Add the headline

Choose the Type tool and click and drag a text box into the upper-right area of the page. Type "Swedish Sashimi*." In the Properties panel, modify the font and size. This example is Gill Sans, Regular, 22 pts. The font has a perfect asterisk for this concept. If Gill Sans is not on your computer, choose another sans-serif typeface.

While the text is selected, choose the Eyedropper tool (it is below the Color Theme tool).

Position the Eyedropper tool over the rice in the foreground. Click on a warm white. The Eyedropper tool will select a color from the image. This technique integrates your text color with the photo. Rather than use a stark white, we've chosen a softer off-white.

When the Type tool and text is selected, the color in the fill box appears as a T and the tiny Affects Text indicator is on.

Tip

The Color Theme tool makes color suggestions based on images in your file. It's fun and easy to use. When you click on an image with the Color Theme tool, the software provides a coordinating color palette. If you want to use these colors, simply click on the + Swatch icon to add the suggested colors to your file's swatches. Access the color Swatches panel by selecting Window > Color > Swatches. Your new Colorful Theme will appear at the bottom of the swatches window. We will use this technique in Chapter 7: Visual Themes.

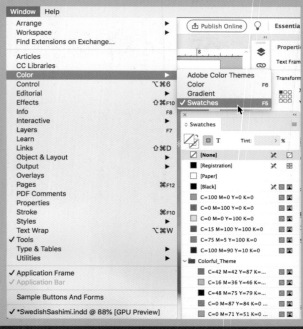

8 **Positioning and locking**

Use the Selection tool and click on the Text box to adjust its position on the page.

Click on the background image (you know it's selected if the blue box is around it) and choose Object > Lock Position.

You can go back and unlock the object at any time using the same menu—choose Unlock Position.

Tip

Sometimes, another frame is selected, and we unintentionally move another object. If this happens, press Command + Z to undo the unwanted move; then lock the background image into place.

9 **Finishing off**

Add supporting copy to the lower-left corner. For good visual hierarchy, we'll use the same font as your headline in a smaller point size. Choose the Type tool, click in the lower-left corner and type "*Food served soft and chewy."

Select the text, choose the Eyedropper tool, and click on the "Swedish Sashimi*" text. Your lower text will be formatted with the headline text box settings. When used with text, the Eyedropper tool copies the typeface attributes (font, size, and color) to the selected copy. In the Properties panel, adjust the size of the lower copy to 12 pts.

Toggle off the appearance of the blue text boxes by pressing Command + Control + H.

Press Command + S to save your work into the "SwedishSashimi" folder.

Press Command + P to print. Use a straight edge and X-Acto knife to trim the page margins to create the full-bleed appearance.

You can share this design as a pdf. Choose File > Export > Format: Adobe PDF (Print).

Nice work.

The advertisement we recreated here was designed by student Kori (Mausner) Mirsberger, with photograph by Mitch Weiss.

Symmetrical Layout | Asymmetrical Layout

Centered layouts are symmetrical; all the elements are arranged equally on both sides of the design. Imagine a vertical line in the middle of this book cover design. The text is half on one side, half on the other, with the type balanced symmetrically along the central axis. Placing the barbed wire across the title creates depth and visual interest, and cleverly integrates type and image.

Symmetrical layouts are a traditional approach. However, centered layouts are not always as visually engaging as this elegant Tatte advertisement. We are drawn to the gorgeous cake, yet, consider, other than the large headline, did you read the copy?

In asymmetrical layouts, designers place dissimilar elements unevenly—yet still achieve balanced results. Though the image may

be more on one side of the page and the headline and supporting copy on the other side, the design has an overall balance. This approach is more challenging than the symmetrical layout, but it is also more visually compelling and therefore more likely to capture and hold your audience's attention.

5.11

5.12

Design Project 4: Path Layout Advertisement

Create your own snack advertisement for the college demographic. What sweet or salty product can you reintroduce to the college crowd using a strong image and succinct copy? For this project, find or take a photograph that has an effective focal point. Come up with a headline. Place the text so that it integrates with the focal point but is not simply on top of it. Add the product website and place this along the path layout.

Students Jamie Balkin and Samantha Lee designed these two projects.

5.13

5.14

The Open House sign and High Line map cover both achieve asymmetrical elegance in different ways. Placed in the top left, the large white Open House text balances the bold colored background. In smaller, simpler fonts, additional information and logos are placed in the negative space on the bottom corners to achieve asymmetrical balance. Can you see how the image of train tracks running through grass creates the focal point in this wide-rectangular layout? And how the semi-opaque white text balances the page?

To achieve asymmetrical balance, consider every position, weight, size, value, color, shape, and texture of each object. A balanced design seems to hold together and feel natural. An unbalanced design makes the viewer uncomfortable. Think about the relative visual weight of all the elements on the page. The position on the page will affect an object's visual weight. What position might make an object appear bigger or more powerful? What position on the page lessens an object's visual strength? This thoughtful approach can be a powerful tool for emphasizing particular aspects of a design.

When creating a layout, place all the elements on the page in such a way as to achieve a relationship based on visual hierarchy and balance. Even if you manage to live well with messy closets, overstuffed drawers, and black-hole backpacks (like me), make it a goal to organize your designs using layout strategies.

Gestalt Theory

Gestalt theory helps us understand human visual perception. The term gestalt means unified whole. It acknowledges that people tend to perceive things as a whole before we see the individual parts. Along with layout strategies, the golden ratio and color theory, gestalt principles underlie many rules for effective graphic design.

Gestalt principles related to graphic design include proximity, similarity, continuity, closure, and figure/ground. When you follow these principles, your designs will appear more coherent and complete. When completing a design, look at proximity—do related lines of text appear near each other? For example, if there are dates for an event, are the times and location in proximity? In the Open House example just discussed, the phone numbers in the sign are grouped. Most designs benefit from the awareness of gestalt theory. We will see examples of the gestalt principle of closure in the next chapter, Logos.

Unity

The gestalt-related principle of unity means that all the elements on a page look like they belong together. Achieving unity becomes more challenging as the number of elements increases. The best approach is to ensure that each object has a relationship with another object on the page. You can do this with alignment, or create visual connections using color, line quality, direction, size, shape, texture, and/or value. The layout itself, whether using a path or grid, can create unity.

In the High Line cover, notice the positions of the white text: both align on the same right-side edge. Additionally, all the text is in one typeface, establishing unity. Variations in size help establish effective visual hierarchy for the copy.

Grid Layouts

From websites to magazines, many graphic designs are multiple pages. Such projects use grid layouts to unify the whole. Turning the pages or clicking menu items reveals the grid layout. Grids are also used for single-page designs. The grid is the underlying structure that allows the designer to create complex pages that look consistent throughout the entire publication. This aids reading, understanding, and enjoyment. Have you ever noticed clever placement of titles or page numbers? That's part of the grid layout.

The word "grid" sounds rigid, yet it does not necessarily constrain the design; rather it allows for many possibilities and experimentation. Its structure, typically invisible, is used as a guide for the placement of images and type on the pages. The underlying grid on web pages helps users access the information intuitively, quickly, and easily.

In these fanned pages of a magazine, all the primary photos are in the same top-left position and their text has the same three-column grid. These are two of the ways a designer achieves unity in a complex document. Can you find others? On the other hand, notice on the last page, there is a fourth column for a smaller photo and a few illustrations that "break the grid," ensuring the design is not too repetitive.

Grid structures are based on the overall page format and an ideal column width of forty-five to seventy-five characters. The character-width guideline is based on the area of a page that readers can see at a typical reading distance without moving their head or getting eyestrain. Three- and two-column layouts adhere to this

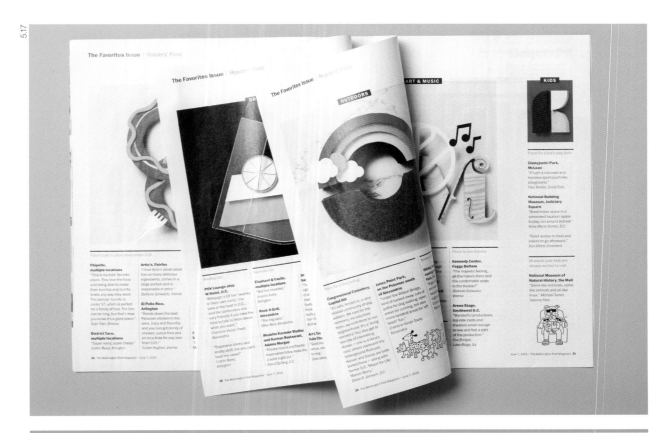

guideline; these are then separated into six or four columns for variation. Many tried-and-true formats exist for standard sizes of screens and papers. Three/six- and two/four-column grid structures are the most common because of their appropriate text-to-column width measurement. Be wary of text columns that are too narrow. It can be difficult to format, or even read, the text.

Standard website formats may have one to twelve columns. When using six to twelve, designers often group columns to create an easy-to-use retail website, such as this Desk Plants design. The online version is below. The top image shows the alignment of images and text to the grid. Notice how the hand and pot photos align to the grid. The empty grid on the right displays the consistent proportions underlying each position.

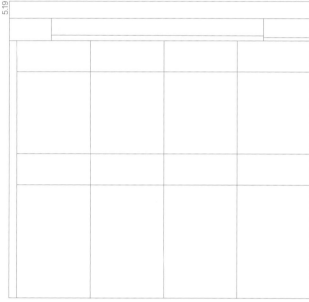

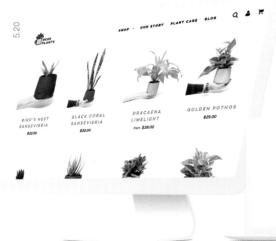

Here is a horizontal grid layout with five columns of text on the left page, and on the facing page, three rows of five images. The published version is on the top. The center image shows the alignment of images and text to the grid. The empty grid on the bottom displays the consistent proportions underlying each position. Notice that the top text aligns with the center of the top row of photos. Each image has a year label consistently positioned in the top-right corner. Where has the designer used the dimensions of a shape as the dimensions of negative space? The repetition of shapes and negative spaces brings order to the overall design.

Here are two spreads from Edible Boston with blue lines indicating the grid. This well-designed grid has repetition and variation, ensuring we intuitively understand the pages are from the same article and making it visually interesting. Every element on these pages aligns with the grid, and the design works on the single-page digital online version as well as these two-page spread formats. How are the images consistent? And how do they vary? Green dominates the five images, with a bit of contrasting brown in three of the photos. There is one wide paragraph for the story, and, for the three recipes, two columns. One is narrow for the list of ingredients. Two fonts are used throughout. More on establishing and carrying out visual themes in Chapter 7.

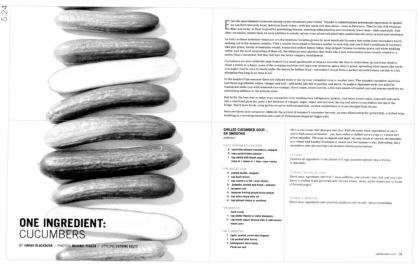

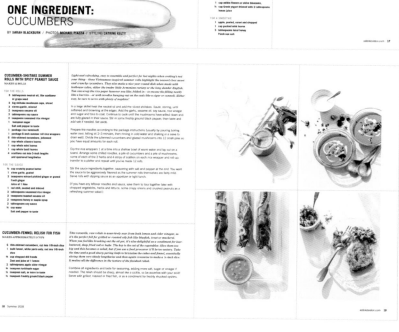

variation. But as grids become too complicated, unity and rhythm are harder to see. Keep in mind that the smaller the squares in the grid, the more choices are required for placement.

Grid inconsistency can lead to lack of visual organization and create confusion. Be consistent throughout the overall design, or you risk losing your audience. Monotony is another risk with a grid structure. That's the reason for a common graphic design phrase: breaking the grid.

Divine Proportions

Grids have a significant history in graphic design that predates text in many parts of the world. Egyptian books of the dead (papyrus scrolls) used grids to organize rows of hieroglyphics. In the mid-twentieth century, designers used this approach to influence international business. After the Second World War, as organizations became more active internationally, designers developed styles that no longer appeared to be from a particular country or culture. Called the

When selecting images for a grid layout, choose images that work together as a whole. The images should have a collectively similar style or palette to be visually consistent throughout a design. This grid is primarily images—how did the designer make fifteen photos work well together and not feel overwhelming? Notice the limited palette. It is primarily greens, a few contrasting warm tones, and even the skies are toned down to appear more neutral than blue. Notice when detailed images are used, and how the figures are cropped to ensure this page tells a complete story of a Boston school's garden.

When choosing a grid, consider the copy. How long is it? How many inserts, subheadings and lists does it have? How many photographs or illustrations? Does your publication or website have more pictures than text or more text than images? In a publication that is mostly text, a simple grid will work well. In a design with many images or charts, a more complex grid allows for greater

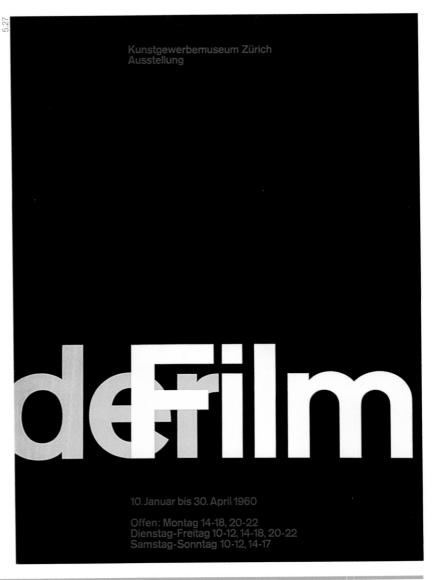

International Typographic Style, practitioners simplified type and imagery and used grid-based layouts.

Josef Müller-Brockmann was one of the most influential developers of the Swiss International Style. He designed the poster "der Film" ("The Film") in 1960. Its proportions are based on the golden ratio, also called the divine proportion. Müller-Brockmann and countless other designers have used the golden ratio to their advantage in graphic designs.

The golden ratio is illustrated with the diagram on the right. Notice the relationship of the rectangle to the square—it remains the same at every size. This is the divine proportion, and it is based on universal mathematical truths. This principle was developed by the ancient Greeks and was considered the most beautifully proportioned rectangle. The Greeks applied the golden ratio in their architecture—the best-known example is the Parthenon.

Using the divinely proportioned rectangle, the "der Film" poster is divided into three columns and five rows. Of the fifteen total rectangles, the top nine approximate a square (the divine proportion). The title fills one of the lower rows, "Film" fills exactly two units of this row. The secondary information on the poster aligns with the "F" of "Film" along the left-hand edge of the second column.

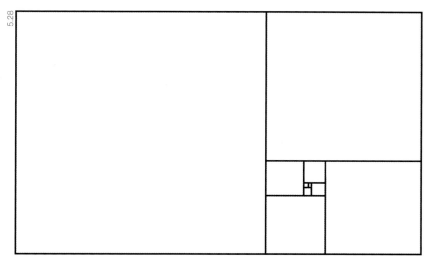

5.28

The design grew out of functional communication needs. The large, white title against a black background projects clearly at great distances. Remarkably, although this poster contains only type, there is an underlying conceptual basis for the text overlap, which suggests the cinematic editing technique of scenes dissolving into one another.

Müller-Brockmann had a huge influence on the field of graphic design. He believed that copy, images, and logos are all subservient to the underlying grid structure in a well-crafted design. In his manifesto *Grid Systems in Graphic Design*, Müller-Brockmann asserts: "The grid system implies the will to systematize, to clarify... the will to cultivate objectivity rather than subjectivity."

It surprises and delights me when I encounter the clear evidence of the contemporary appeal of golden ratio principles: from website designs, to wine bottle labels, to the tattoo on the shoulder of an architect I know.

Design + Software Skills 5.2:
InDesign Grid Layout

Objective
Create a grid layout and place text from a Word file into a frame. This project also reinforces the type tips from Chapter 3 and provides an opportunity to create an asymmetrical layout.

To Do
Create a grid layout for the Type Tips text.

Tip

InDesign allows us to place very large Microsoft Word text documents into a file. When the amount of copy is more than one frame or column, hold the Shift key down and the copy will automatically flow or thread through as many frames as needed. For this exercise, we'll use three frames for the heading, subheads, and copy, and no threading is required.

1 **Make the grid**

Open a new document in InDesign. Make it letter size, one column and 0.5 inch (1.27 cm) margins on all sides.

Draw the grid. Press Command + R to make rulers appear at the top and left of the file. Place your cursor in the left ruler, click and drag a vertical guide to the 1 inch (2.54 cm) mark. Then draw three more vertical guides at the 2.75 inch (6.99 cm), and 3 inch (7.62 cm) points on the page.

Place your cursor in the top ruler, click and drag to place horizontal guides at the 1.5 inch (3.81 cm), 2 inch (5.08 cm), 2.75 inch (6.99 cm), and 9 inch (22.86 cm) points. You've drawn the grid.

2 Add frames

Draw three frames for the copy. Select the Frame tool. Click your cursor at the first intersection of the top two lines. Drag it to the next intersection of lines. This will frame your title. The three x-ed boxes on the right show the frame positions.

For your subtitle column, draw a frame that fills the area between 1 inch (2.54 cm) and 2.75 inches (6.99 cm), horizontally, and between 2.75 inches (6.99 cm) and 9 inches (22.86 cm), vertically.

For the Type Tips copy, draw a frame that starts at the 3 inch (7.62 cm) point and ends at the 8 inch (20.32 cm) margin, horizontally, and 2.75 inches (6.99 cm) to 9 inches (22.86 cm), vertically.

3 Add copy

Add the copy. Select the Type tool, click in the top left frame, and write "Type Tips." Format the text. This example is Impact, 30 pts, right aligned. Impact is bold yet narrow. If it's not on your computer, try another bold, condensed, sans-serif font.

4 Adjust the alignment

Modify the position of the words in the frame. Currently, they align to the top of the frame. We want the grid to be apparent, so we'll change the position to bottom.

Choose Objects > Text Frame Options. In the Vertical Justification section, change the Align setting to Bottom. Press OK. Now your title sits at the lower edge of the frame.

5 Make a new folder

Create a new folder on the desktop for this project. Click on the desktop to make it active and choose File > New Folder. Click on the untitled folder and type 'Grid'.

6 Acquire the text file

To acquire the copy, download "TypeTips.doc" from http://www.bloomsbury.com/graphic-design-essentials-9781350075047 to your desktop. You should see a thumbnail image on your desktop called "TypeTips.doc."

Click and drag "TypeTips.doc" over the Grid icon to put it in the folder. Although InDesign does not create a link to the Word document, it's a good way to keep track of all the files for each project.

7

Select the Type tool and click in the largest frame. Choose File > Place and choose the desktop Grid folder, then "TypeTips.doc" from the file listing. Press Open. The text is placed in the frame.

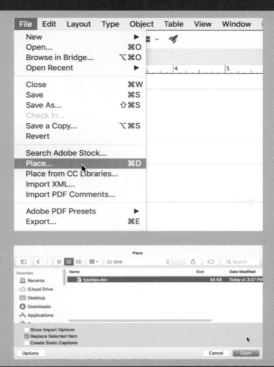

8

Format the copy

Select all the copy and format. While the Type tool is selected and positioned in the large frame, press Command + A to select all. Use the Properties panel to format your text. This selection is Cochin, 11 pts, Regular, left-aligned. Cochin is a serif typeface. If it is not on your computer, try another serif font that contrasts with your title.

Tip

If you can't find the Properties panel to format your text, simply reset the workspace by choosing Window > Workspace > Reset Essentials.

9 Add subtitles

Finally, add the subtitle text. Using the Type tool, type "Typeface" in the top of the empty frame. Highlight the copy and format it. This selection is Impact, 11 pts, right aligned. Use the same font as your title and make it smaller. Once "Typeface" is formatted, press Return so that the cursor moves down the column adjacent to the next group of tips and type "Contrast"; then press Return several more times to bring the cursor next to the third group of tips and type "Layout." If the lines of text don't align, check that the leading is the same for both columns.

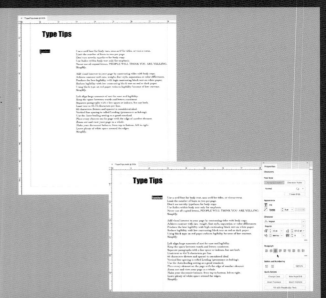

10 Outline the title

Modify the title by highlighting Type Tips and pressing the color bent arrow. This switches the fill color to the stroke color and creates an outline text appearance. In the Properties panel, change the Stroke weight to .5 pts. Save your work, then try different fonts, colors, and placement.

11 Finishing off

View your work without the grids by pressing Command + ;. Then press Command + Ctrl + H to hide the frame edges. Or, when nothing is selected, you can press W for a quick way to view your work without the frames and guides. Press W again to see the guides and frames.

Press Command + S and save the file to your Grid folder.

Press Command + P to print.

Press Command + E to export as a PDF.

Nice work!

Major Points Summary

— Layouts are used to organize content and enhance the communication of information.

— Contrast is often used to create good focal points.

— Path layouts begin with a focal point, the area of emphasis in the design. All the supporting elements are positioned to lead the eye through the entire page.

— Integrate the type with the image to ensure that the audience reads the copy.

— Symmetrical layouts align all the content along the center axis of a page.

— Asymmetrical layouts consider the position, weight, size, color, and shape of each element to achieve an overall balanced page that is not aligned on the center axis.

— Gestalt means unified whole. People tend to perceive things as a whole before we see the individual parts.

— Unity is a principle of design that means all the elements look like they go together.

— Multiple-page designs, such as websites, magazines, and catalogs, have grid layouts.

— Grids are invisible underlying structures on which images and text are positioned.

— Grids establish and maintain visual cohesiveness throughout a publication.

— Grids help your audience read and understand content.

— The ideal column of text is forty-five to seventy-five characters wide.

— Standard paper formats and ideal column width lead to typical three/six- and two/four-column grids.

— Standard website formats can have one to twelve columns. Often columns are grouped to three or four.

— Müller-Brockmann used grids based on the divine proportion (golden ratio). He was a prolific designer, whose work is still exhibited, and a writer, whose books remain in print decades after they were written.

— The ancient Greeks developed the golden ratio; it is considered the most aesthetically pleasing rectangle. They based their architecture on this proportion.

Software Skills Summary

— InDesign introduction, image, and text frames.

— Overview of the toolbox and the most commonly used tools: Selection, Frame, Rotate, Gradient, Direct Selection, Type, Shape, Scale, Eyedropper, Zoom, Fill color and Stroke colors, Color Toggle, and Formatting Affects Frame and Formatting Affects Text buttons.

— Keyboard shortcuts.

— InDesign skills: place images, full-bleed, document formatting, frames, Direct Selection tool to transform or move an image within a frame, Color Theme tool to make palettes, Eyedropper tool to select colors, swatches, lock position of frame, place text from Word documents, thread text, draw guides for a grid, modify text frame options, vertical justification of type, format text, Eyedropper tool to format text, outlining text, viewing frame edges, and guides.

Recommended Readings

For additional instruction on grids, read the master's book: Josef Müller-Brockmann's *Grid Systems in Graphic Design*. Also check out the thoroughly enjoyable *Making and Breaking the Grid* by Timothy Samara.

Logos

Paul Rand, Eye Bee M poster, 1981

Created for a company event by Paul
Rand, this poster was temporarily
banned at IBM for fear that employees
would take liberties with the logo. This
poster is a rebus where two letters are
replaced with pictures for their sounds;
it is typical of Rand's playful approach
to design. The IBM logo is based on the
City Medium font; horizontal lines were
added later to unify the letters and evoke
the lines on a computer screen. Rand's
ability to develop innovative concepts
and reduce the message to its symbolic
essence made him influential while still
in his twenties. Later his work with Bill
Bernbach became the inspiration for the
copywriter/art director partnering that
was adopted by the advertising industry.
Rand taught several generations of
designers at Yale, including some of
my professors (one of them told me
his critiques made him cry).

6.01

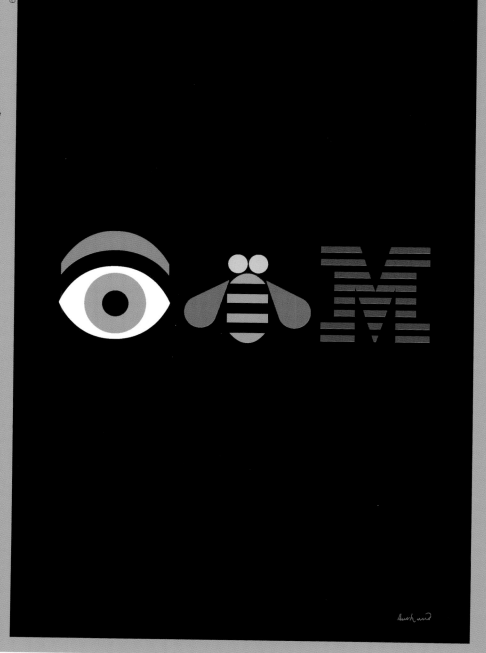

Simple but Complex

Customized typography, meaningful color and refined symbols are common features of successful marks. These are ideation sketches for Fortify, a company that produces exceptionally strong 3D prints. Triangles are the strongest shape, so the designers rotated the F to form a triangle and express the product's strength and stability.

Logos express complex company personalities with very simple yet distinct designs. It's challenging and rewarding work. Read on for strategies to create effective logos.

Logos are remarkable—these simple little marks communicate an enormous amount of information. They identify products or organizations, convey quality, style, value, even origin. When you see a BMW car, do you immediately know where it was made, its approximate cost, and reliability? How does this assessment differ from one you might make about a car bearing the VW logo? Opinions will vary somewhat, but both symbols are understood immediately. Logos can be an endorsement of excellence, and they can add value to a product. Will you pay more (or less) for a pair of jeans with a noticeable logo?

Successful Logos

Logo design is challenging: express a multifaceted message with a minimal yet unique mark. Logos must be visually uncomplicated for three reasons: ease of recognition, recall, and reproduction.

Recognition | The best logos are so simple that they are recognizable even when there is only a quick or partial view of it. Often, we see a glimpse of a logo while scrolling online or a view obscured by a physical object. Do you need to see all the letters in Coca-Cola before you grab a can?

Recall | An intricate mark has too many details for us to remember. A simple mark is more easily recollected. I'll bet you could sketch the Nike swoosh or McDonald's M accurately from memory. Try it.

Reproduction | Logos must be reproduced in a variety of media and sizes, from websites to illuminated signs atop buildings. Logos should appear as crisp and clear at less than an inch as they appear when 25 feet in diameter.

American Express is recognizable on all sorts of media—from apps to football jerseys, and it is considered one of the world's most trusted brands. The logo was recently refreshed to optimize its performance across various platforms. The design challenge was to strengthen a well-known brand for digital products. In smaller digital displays, it can be difficult to make visual impact if the mark is too detailed. Some elements were simplified, and others appear bolder. The results are a seemingly more confident brand.

See the progression of American Express logo styles below. Each version made the logo clearer and stronger with simplification of color, typography, and the form of the symbols. Notice the square logotype replaced a gradient with a new flat blue, simplifying the color. Take a close look at the smart simplification of the centurion. These brand signifiers have been carefully redrawn to be more digitally legible and reproducible.

The refresh includes lettering that was developed to express the spirit of the original logo. The outlined lettering was redrawn with a non-outline version that appears outside of the blue box, as on the jersey. Outlined fonts often have legibility issues in smaller digital displays. The logotype was custom drawn in various scales for visual continuity and legibility at different sizes and media.

Usually, the number of colors in a logo is limited to one or two because each additional color adds to the printing costs of basics such as business cards and stationery. Additional colors also take away from the simplicity of the mark. Logos also need to reproduce well in black and white. When even desktop printers use color ink, why do we still need gray scale versions? Low-cost publications are still typically printed with black ink. And, often we make logos in one color when grouped on websites or apps, such as social media icons. It's an additional challenge to design a multicolored mark in black and white, but it's necessary for the variety of media we work with.

Think of logo design as providing a complex description in a single visual statement. To be competitive in the marketplace, organizations are often diversified—they provide a variety of products, or more than one type of service. When creating a logo, designers cannot emphasize one particular aspect of a company without regard to all the others, or those parts of the business that were not featured would be poorly served. This could have negative consequences for sales, not to mention causing nasty battles in the conference room.

The National Geographic Society was once primarily known for the magazine that had a traditional masthead and a bold yellow border on the cover. But now, Nat Geo's products include documentaries and television channels. To establish a cohesive branding policy for a vast range of products, the design team created a rectangular mark based on the magazine's iconic border. Now the brand is recognized by this simple yellow rectangle on products ranging from websites to puffer jackets.

6.04

6.05

6.06

6.07

6.08

6.09

Types of Logos

Logotypes

There are several styles of logos. The classic is the logotype, which is defined as the name of the organization presented in unique typography. The mark can depict an entire name, an abbreviation, or initials. Coca-Cola has the most recognized logotype in the world. Their design is so well-known that we easily make it out even when it appears in different alphabets such as Arabic or Chinese.

If you are a coffee lover, George Howell could be your hero. For three decades he has travelled the world discovering and roasting the highest quality coffee beans possible, and he devised ways for coffee farmers to benefit financially. In many ways he is an international coffee champion, and appropriately his logo is based on his signature—just black, no milk.

The typography and color accurately convey the style of 1970s psychedelic rock band Väg. You'll see the band's album cover designs in Chapter 7.

Initials

Logotypes sometimes feature initials of the company name to create more easily recognizable identities. Aslan Foundation works to preserve and enhance liveability and culture in Knoxville, Tennessee. At first glance the initial seems to be an elegant capital A, yet the logotype is made of geometric forms, making it distinctive and consequently more memorable.

Halstead Real Estate's logo is a multidimensional H. The icon's H is appropriately architectural. It can stand alone, and its open form offers the flexibility for use in patterns, as a frame for photography, and, when rotated, as a directional arrow. Halstead uses a variety of palettes that reflect a growing movement in design that redefines the role of color in corporate branding. Alongside strong visual elements, the use of color allows a brand to appeal to audiences by tapping into personal connections that people may have to things like seasons.

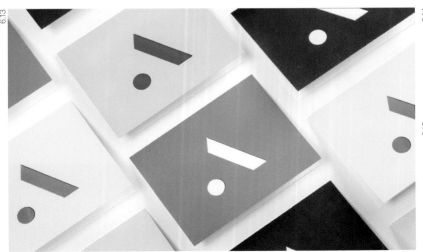

Abstract Symbols

Logos often combine typography with symbols. Abstract symbols are common because no particular area of an organization is favoured over another. Some abstract symbols have characteristics that allude to the general nature of the business. When there is a conceptual basis for the abstract shape, such as with Nike, we call that an allusive-abstract symbol. Nike's logo is an allusive-abstract symbol with a dynamic quality that suggests movement—perfect for this brand of athletic gear.

Tom Geismar, Chase Manhattan Bank logo, 1961

This logo was created in 1961 when two banks merged to form Chase Manhattan bank. This is a purely abstract symbol that was rare at the time and the first for a bank. "It's not us," said one of the executives who expected to show off their new sixty-story building on the letterhead. But six months later, his logo-patterned tie had become a favorite. Radical for its time, the abstract Chase symbol has survived a number of subsequent mergers and has become one of the world's most recognizable trademarks (and proof that sometimes the designer knows best).

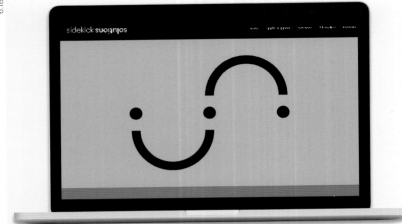

Pictorial Symbols

Pictorial symbols are representational. They look like the product or have an obvious association with the company's business. Originally, Animal Planet's logo was an elephant with a round graphic of earth at the tip of its trunk. The channel needed an updated logo so it could reproduce well on a variety of digital and physical products, as well as be recognized globally. Elephants have a majestic character that enables them to represent the entire animal kingdom, and here its running legs create the arc of planet earth. Does anyone else see a breaching whale in the head and trunk shape? Logos that have subtle details that may have multiple interpretations resonate with audiences.

The US Open's previous logo included a flaming tennis ball that was challenging to reproduce consistently, especially in digital media. That earlier mark did not convincingly convey a premium sporting and entertainment brand. Additionally, the brand was less recognizable because there were many versions. By simplifying the symbol, the concept of a flaming

Simple geometric shapes in abstract symbols can represent a wide array of subjects. Two curved lines and three dots combine to form a unique S for Sidekick Solutions, a technology support company. Can you see a smiling face suggesting Sidekick solved the problem? What else can you see in these abstract forms?

The logo design of BLOC coffee café is based on the concept of tangrams: geometric shapes combined to form the letters of the name. Notice the coffee bean (the O) in the logotype. The same shapes are playfully recombined into espresso pots and steaming mugs.

N O R O S H I

F O R K L I F T

C A T E R I N G

tennis ball can be used to convey the excitement of this premier tennis tournament. This is also an example of how we are wise to choose colors that communicate information in our logo designs. The new mark simplifies the symbol, color and typography to convey both the ball and flame. This logo is highly legible and its white sans-serif font uses the same letter for both the u and n.

Noroshi is a ramen restaurant originally from the historical port city Hakodate, in the beloved Hokkaido region of Japan. The characters for Noroshi mean smoke signals, which is cleverly illustrated with this elegant arrangement formation of the characters のろし.

Only sandwiches are served at Mike & Patty's Boston, and they are the city's best. Here, a slice of bread makes a playful enclosure shape. The pig is the featured associative symbol because you can get bacon on any of their eight menu offerings (and you should!).

Do you remember our discussion of the *Survivor* logos in Chapter One? This design uses some of the same strategies, including good visual hierarchy of text and the use of color to convey information. The logo colors appropriately suggest hearty flavors and sustainability, reflecting the micro-locality of their ingredients.

Associative symbols depict images that are not the product or service itself, but ones that have a strong relationship to the organization. Forklift is a caterer for special events. The company is easily identified by a serif logotype that is balanced with a cleverly drawn fork. A fork illustration is neutral, it does not emphasize a specific cuisine, and it makes an effective associative-pictorial icon for their brand. When we look at this logo, we immediately see a fork, but when you look closer, this icon is made of four curved lines that don't intersect. These four curved lines read as a complete object. This is due to the gestalt principle of closure, which recognizes that our eyes will automatically complete figures even when part of the information is missing.

BATTERSEA
HERE FOR EVERY DOG AND CAT

The logo for Battersea Dogs and Cats Home is impressive on a couple of levels. It is funny, and a unique and beautiful use of watercolor. Using a variety of hand-drawn symbols ignores many traditional logo symbol design strategies, and yet it is very successful.

Battersea, known by its first name, was established in 1860 to care for cats and dogs. Since I learned of it, I hear references to "Battersea" in British culture frequently. This is a beloved institution but one that needed a consistent visual identity. Charity organizations can tend towards overly sentimental forms of communication, but the design team wanted an honest and straight forward voice for the new branding.

This variety of heads is used to brand a wide range of services and represent the motto, "Here for every dog and cat." Their faces don't have features, yet they remain expressive with a sense of individuality, and this approach avoids depicting the animals as victims.

The Franklin Gothic font injects authority into the identity. The charming dogs and cats, together with the strong text, enable the Battersea visual identity to convey both compassionate care and animal welfare expertise. This identity system can flex and adapt based on the audience.

Flexible Systems

Brands often have a great variety of products, services, and locations. Flexible identity systems allow clients to adapt to these variations and for company growth. Shake Shack's flexible system helped build it into an upscale, casual-dining destination internationally. Now on five continents, this restaurant started as a hotdog cart in historic Madison Square park. Originally, the cart was to help support the refurbishment of this park in Manhattan, yet the hotdogs and burgers were so popular, it grew into a permanent kiosk. That first structure's architecture was inspired by burger drive-in restaurants from the 1950s and 1960s. Now, the storefront signs are always in midcentury-style font—Neutraface.

Shake Shack's flexible system uses fonts that seem to be both retro and contemporary, along with neon-style graphics. The system includes a script and sans-serif Futura on the menus, packaging, and swag. The brand's colors are neon green, white, and black, and packaging is simple, elegant yet playful. The iconic graphics create a welcoming and fun brand experience.

6.28

6.29

Simon & The Whale is another flexible system for a New York restaurant. This restaurateur's son's love of whales inspired the name and logo. There are three distinct characteristics of this logo: a rotated ampersand, retro colors and a handmade texture. All combine to create a unique and appealing brand.

Illustrator software was used to mimic the look of etching and achieves a handmade look. Four 1960s-inspired color combinations and logotypes for both the full name and initials make for a flexible brand where there are multiple, yet distinct, versions of the mark. For example, the different color combinations are used on menus to indicate breakfast, lunch, or dinner; and the S&W is a tertiary mark that can be used in conjunction with the primary logo or on its own. These marks can then be used interchangeably to give a more nuanced and layered brand experience. Read more about establishing visual themes for flexible systems in Chapter 7.

SIMON & THE WHALE

SIMON & THE WHALE

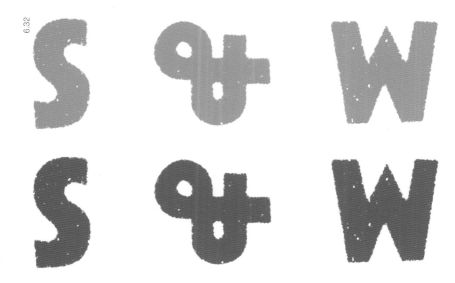

Semiotics

Semiotics in graphic design is concerned with the use of signs and symbols for communication. These are used for many different purposes in graphic design. The Sidekick Solutions brand is a flexible system that demonstrates the use of semiotics in branding. Consider the specific meanings of these common symbols, arrows, and caution signs, when used by this technology support company.

Function Engineering's expertise is hinge and linkage mechanisms for consumer electronics, including gaming headsets, robots, and phones (wonder if you're using one now?).

The logo and a series of icons are made of type forms that mimic hinge and pivot systems. Look at these type forms individually and recognize how these symbols are abstract and essentially semiotic. The flexible system of icons allows for illustrations and patterns that can be used for a variety of collateral across all media. Notice the use of negative space that defines the objects. When printed, this system requires only black ink on white paper, which has the added advantage of lower costs.

6.33

6.34

6.35

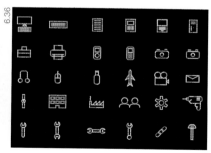

6.36

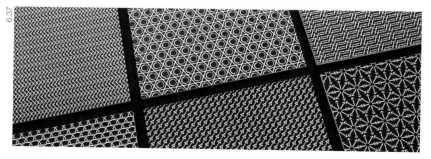

6.37

Design Analysis:
Logotypes

LOGO
TYPES

Shake Shack

LOGOTYPE AND PICTORIAL SYMBOL

INITIALS AND ALLUSIVE SYMBOL

COCHON
DINGUE

LOGOTYPE AND PICTORIAL SYMBOL

ABSTRACT SYMBOL

Objective

This is an opportunity to analyze the success of existing identities and reinforce your knowledge of underlying strategies for logo design. This is also a page-layout design opportunity.

To Do

Find four examples of different types of logos: logotype, initials, abstract symbol, and pictorial icon. Place all the examples on a page and label with the type of logo.

You can create this analysis with any of the software, but ideally it will be done in Illustrator. Open a letter-size file in Illustrator. Search the internet for good quality images of logos. Click and drag each logo to copy into a folder on your desktop. Place the logos into the Illustrator file. Make all four logos roughly the same dimension, approximately 1 to 1.5 inches tall. Remember, don't enlarge an image from the internet; its size has already been compressed. Now label each logo and design the page keeping in mind the type and layout tips from Chapters 3 and 5.

Tip

Notice how the typefaces and colors suggest each brand's personality. Analyze whether the symbols are purely abstract, allusive-abstract, pictorial, or associative-pictorial. Notice how the object is stylized or simplified in the symbol.

The Creative Process

How do you take on the daunting challenge of logo design and achieve a rewarding solution? Successful designers follow a creative process when tackling projects—these are the steps they take to come up with fresh ideas, and then they develop them into flawless final results. Everyone eventually develops a unique creative process; you will too.

These five steps are suggestions for new designers to ensure successful projects—with less frustration—at this stage in their careers.

The Creative Process

1. Understand the problem
2. Get inspired
3. Brainstorm
4. Sketch
5. Produce

1. Understand the problem. Restate the design project in your own words. State what the design is intended to communicate and to whom. This is valuable practice for the professional world, where you will reiterate your understanding of the project to the client, ensuring you understand expectations before you begin the work. In professional studios, detailed reports of this kind are called design briefs.

Research your client and their industry. Understand their organization, products, services and competitors. This knowledge will feed your creative ideas.

2. Get inspired. Expose yourself to art. Inspiration can happen on the subway, but it's far more likely in a gallery. Make casual visits to art galleries and museums. There may not be a direct relationship between what you see in an exhibit and the project you're currently tackling. You may be inspired by artists' use of design elements: color, direction, line, shape, size, texture, and value. Use this inspiration to inform your creative ideas and projects in unexpected ways.

3. Brainstorm. Gather a couple of friends and brainstorm about the project. The goal is to generate fresh ideas and associations, so have fun and don't hold back any thoughts. There is only one rule about brainstorming: no idea is a bad idea. Encourage unrestrained contributions and don't criticize any input during the session. After the session, edit the ideas down to the most promising ones for your project. This will give you a few different approaches to pursue.

4. Sketch. Get into the habit of sketching ideas before your fingers touch the keyboard. Everyone's ideas look fabulous in their head, but seeing a draft on paper is a way to determine if a concept will be successful before you spend time producing it with software.

Use thumbnail sketches to work out your layouts. They're small and simple—elaborate, time-consuming illustrations aren't necessary or practical. Start with one page of nine to twelve rectangles, using the same proportions as the final design. Think with the marker in your hand. Don't stop to erase or add color. It's a way to think visually and quickly develop layouts. You'll know immediately from the thumbnail sketch if a design is worth exploring in depth. Once you have a couple of ideas you're happy with, choose the best, make a larger sketch, and add details. This will save you valuable time on the computer.

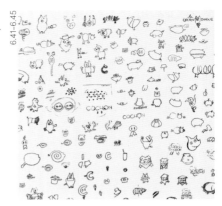

When you have a complete design, step back and evaluate your work. Ask for feedback. If you can, allow some time to pass, then come back and take a fresh look. If it will be printed, look at printed versions. Make the necessary revisions. Before presenting to a client or class, rehearse a brief explanation of your concept and choices.

Follow this five-step creative process with your design projects, and you'll have an understanding of your subject, fresh ideas, and a blueprint for your design. When you switch to the computer, you'll move in a definite direction. When you present, it will be the best work you can produce.

Cochin Dingue means crazy pig in French. Notice the number of sketches made for this beloved 35-year-old restaurant. The brand needed a clear and concise update that remained authentic and Québécois. The owners debated the appeal of these 182 sketches. Sketches are not complete illustrations and do not need formal details to help you and your team create. As you see here, sketches are simple and one color, but they enable teams to think visually and collectively before spending time developing the final version.

The client also wanted to emphasize their French flair and enliven their brand with bold colors. Eventually they chose an active pig that also suggests the French flag. Historical French poster art inspired the choice of fonts for this award-winning brand design.

Don't skip sketching! The Citibank logo was sketched during the initial client conference (by Paula Scher), and now it's called the $1.5 million napkin. The red arc over the t is a nod to the umbrella that was part of the original Traveler's logo when Travelers Group merged with Citicorp and started Citibank. It is advantageous to have this abstract arc, and the curved shape is used throughout their brand, including arcing of architectural details at banks and ATM kiosks.

5. **Produce**. Once you've got a clear idea from your sketches, begin to produce the work with software. Use Illustrator for vector illustrations, logos, and single-page designs. Most logos are designed in Illustrator. Use InDesign for multiple-page layouts and prepress. Use Photoshop for optimizing photographs and creating complex illustrations. Designers often combine work in all three for their projects.

Designing Logotypes

When creating a logo, you'll choose a font and color scheme, and sometimes add a symbol. All these decisions combine to identify and distinguish the company from others. An essential aspect of logo design is taking the time to analyze typefaces. Choose a font that expresses the personality of the brand; then, if necessary, you can customize the letters in the name and choose colors that set the tone.

Type the company name with all capital letters, all lowercase, and upper- and lowercase. Try each typeface in serif, sans serif, and depending on the client, include novelty fonts. This selection of fonts features sans serifs. Look closely at the characteristics of the letters. There are many variations of the capital M and A; some are pointy, while others have flat tops. It's surprising, once we take the time to look, how many letters have distinctive forms. Use these characteristics to give personality to the mark. From top to bottom these typefaces are Futura, Gill Sans, Century Gothic, Bank Gothic, and Franklin Gothic.

In the citi logo on the previous page, all are lowercase letters of the same height except the lowercase t, which forms the umbrella shaft and handle.

Have you ever noticed the variety of dots used for lowercase *i*s? Sometimes they are ovals or squares, and the font for Sidekick Solutions has perfect little circles (see page 171). This detail inspired the sideways S 'smile' we discussed earlier in the chapter. Additionally, the repetitive angles of the *K*s create a dynamic, fast quality for the logotype.

Notice the relationships of the letters within the name, look at the white space created by the letter spacing—is it open or enclosed? Have you ever noticed the evocative shape between two letters of the FedEx logo? Look between the E and the X for a symbol that represents the nature of the delivery business. This logo required a customization of two fonts, Univers and Futura, to get the arrow just right.

Notice the Bar Sardine logo, the ascender of the lowercase b balances the elongated S nicely. The ascender of the lowercase d in "sardine" pulls the two words together. This logo was painted by hand, resulting in the distinctive watercolor texture. Next it was scanned and adjusted in Photoshop.

Rhythm can be established within a word by the pattern of curves and lines created by the letterforms. Is the overall shape of the name a strong, imposing all-caps rectangle as with FedEx and George Washington Bar? Or do an irregular baseline and swirls add vitality, as in this GW Bar mark? Think of the applications of a primary mark that uses all capital letters for the full logotype, and how would you use the secondary initials in the logo?

MARK	Mark	mark
MARK	Mark	mark
MARK	Mark	mark
MARK	MARK	MARK
MARK	Mark	mark

GEORGE
WASHINGTON
BAR

GW
Bar

Designing Symbols

You can develop a design further by combining the name with symbols. A key to a good logo symbol is simplicity of form. If you want a pictographic icon, strike a balance between characteristic details and streamlined form. Determine which features of the object should be emphasized in the drawing. Mamaleh's all-caps name fits the curve of the circle enclosure while framing the cute and linear face icon. The circular icon integrates with the logotype and stands comfortably on its own. The face icon emphasizes hair and reduces the facial features by using dots for the eyes and tiny curved lines for the nose and mouth.

Sports team logos are a good source of inspiration for symbol drawing—notice how designers emphasize features that express competitive characteristics. The stride and flowing mane of a horse show speed, while wide sharp horns suggest a bull's strength and power.

The black-and-white markings of a Boston terrier are emphasized in the Boston University Terrier logo. His crossed arms, spiked collar, pronounced teeth, and erect ears suggest intensely competitive teams. The school's signature red is used for the collar and inner ears.

Always look at an example of the object while sketching—don't trust your visual memory for details. Emphasize one or two details and simplify the rest. George Washington's image is on the coins I carry in my pocket, yet I wouldn't draw his portrait without looking closely again. A painting in the client's bar inspired this cameo style image of George Washington. The contrasting color portraits have a fun detail that is activated when these two are an animated GIF. Can you find it? George winks when the two GIFs are animated.

Try several different colors and combinations. Refer to the color connotations chart in the back of the book for guidance. Go beyond your personal favorites and select colors that are appropriate for the client. Lemon yellow works perfectly for Desk Plants, but it may not be advantageous on a law firm's letterhead. What colors would be suitable for a law firm? Use color to convey information about your company's brand.

Desk Plants is an internet start-up with a mission to brighten your workspace with live plants. They sell several types of plants, and this is suggested in the variety of greens in the leaves. The symbol also conveys the size of the plant and container and even implies an affordable price.

Finally, make a black-and-white version that can be used for one-color and low-cost printing requirements. The logo may need some minor modifications to lines and fills when converting.

Practice a client presentation for your logo in front of a friend or a mirror. Explain the decisions you made during the development—why you chose the font, colors, and symbols to represent the brand. Clients are more likely to appreciate the design results when they hear how their company was thoughtfully interpreted.

Design + Software Skills 6:
Illustrator Convert Type to Outlines, Redraw Letters, and Place Symbols

Objectives
Develop Illustrator type and drawing skills while learning how to customize letters for logos.

To Do
Recreate a logo. Create outlines from text and customize letters. Draw a jagged edge on the R and place the climbing figure. Use Pantone spot colors for the logo.

1 **Format the name**

Start Illustrator. Open a new file that is 7 inches by 5 inches (17.78 by 12.70 cm).

Type "ROCK" and make the font Industry, Ultra, 120 pt. You can get this font from Adobe Fonts by typing "Industry Ultra" into the font window and clicking on the Activate icon. If you don't use Creative Cloud, try another ultra-bold sans-serif font.

2 **Create outlines**

Convert the text to shapes. Confirm the name is selected and choose Type > Create Outlines. Ungroup the letters by clicking the Ungroup button in the Quick Actions panel. Or, use the keyboard shortcut Shift + Command + G.

3

Customize the R
Select the white Direct Selection tool.

4

Make the curve a straight edge
Click on the anchor point (small white square) on the lower right of the curve. When selected, it will change from white to blue. Then click and drag the circle at the end of the direction handle to make the line straight. Drag the direction handle to the top anchor point.

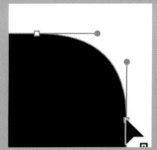

5

Repeat with upper point
Click on the upper anchor point. Then click and drag the circle of this direction handle to make the line straight by dragging the circle to the bottom anchor point.

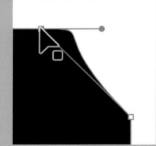
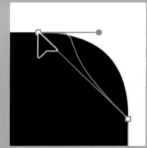

6

Repeat with the lower curve
Zoom to the lower curve. Click on the lower anchor point. Then click and drag the circle at the end of the direction handle to make the line straight. Drag the direction handle to the top anchor point.

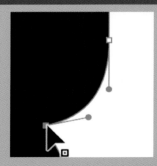

7 Repeat with upper point

Click on the upper anchor point. Then click and drag the circle at the end of the direction handle to make the line straight. Drag the direction handle to the bottom anchor point. Phew! Well done!

 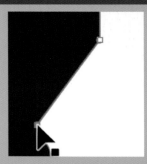

8 Gorgeous!

Optional: You can continue to make the O and C curves straight for extra practice or go to the next step and leave the O and C curves as is.

9 Lock the R

Select the R and lock it into position, so it doesn't move as you draw the jagged line. Click on the R, then choose Object > Lock > Selection (or use the keyboard shortcut Command + 2).

Object	Type	Select	Effect	View	Window	Help

ROCK

Transform ▶
Arrange ▶

Group ⌘G
Ungroup ⇧⌘G
Lock ▶   Selection ⌘2
Unlock All ⌥⌘2   All Artwork Above
Hide ▶   Other Layers
Show All ⌥⌘3

Expand...
Expand Appearance
Crop Image
Rasterize...
Create Gradient Mesh...
Create Object Mosaic...
Flatten Transparency...

Make Pixel Perfect

Slice ▶

10 Customize the R

Select the pen tool. Click the bent arrow toggle to make the fill color no fill. Then double-click on the black stroke box to open the Color Picker window and choose magenta for the stroke color. Magenta is a temporary color that enables you to see the line against the black R.

11 Draw the jagged line

Begin just below the top left edge of the R. Click, drag, and click to make the first short diagonal line. Click lower and slightly to the left for the next line, then again slightly lower and towards the right. Continue down the left side of the R. Notice midway down, there is a straight line where there's a hold in the wall. Continue a jagged line to the bottom edge of the letter.

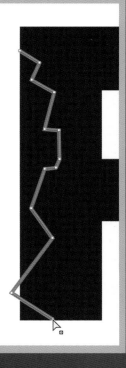

12 Close the path

Make a complete shape by clicking to the left, make a vertical straight line towards the top of the R, and finally click on the very first point to make a closed path.

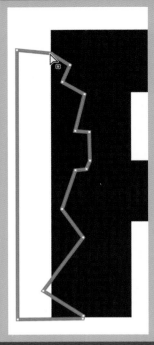

13 Make it white

Make the closed path white by clicking on the shape, selecting white fill, and no fill for the stroke. That looks cool, doesn't it?

14 Place the figure

Download the climber figure from http://www.bloomsbury.com/graphic-design-essentials-9781350075047. The filename is RockClimber.ai. Then place it in your project folder.

Select File > Place to copy the figure into your Illustrator file. Drag the figure into position on the wall.

15 Scale the figure

Select the Scale tool. Click on the figure, hold the Shift key as you scale the figure. The top of the figure's line will align with the top of the R, and the bottom of the line will align with the bottom of the R.

16 Apply Pantone color

Select Object > Unlock All. Next, make the ROCK letters red, specifically Pantone color 186 C. From the bottom of the Windows menu choose Swatch Libraries > Color Books > PANTONE+ Solid Coated. In the Pantone window, type 186 into the search box, then click on the line that says PANTONE 186 C to make the ROCK letters red.

Tip

When using Pantone color swatches, choose "Small List View" (in the hamburger menu) to see each color's number.

17. Add Text

Using the Text tool, add CLIMBING, in all caps, Industry Ultra, 36 pts, and make it right aligned. Align CLIMBING to the right edge of ROCK. Add Boston University to the top left of the name and make it Avenir, Book, 18 pts. Align Boston University to the top left edge of ROCK.

18. Change the text colors

Using the Selection tool, click on the text box for CLIMBING; hold the Shift key down and click on the text box Boston University. While both text boxes are selected, choose the Eyedropper tool. With the Eyedropper, click on the climber figure so the text changes to blue.

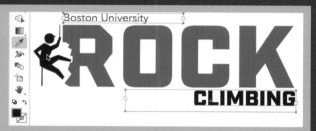

19. You did it!

Well done! You can use these software skills to make your own logos.

The logo we recreated here was designed by student Darah Rifkin.

Design Project 5: Logo Design

Design a logo. Use only two colors and add a tag line. Choose one from this list of fictitious companies or design a logo for an intramural sports team at your school.

Toro – a tapas restaurant
Zeus – a Greek restaurant
QK – a GPS app
Circle – a restaurant
Memo – a paper store
Intermural Sports Team

These logos were created by students (from the top left to bottom right): Yuyang Yuan, Yunyi He, Ryan Huff, Heather Hamacek, Cara Kingsley, Xinyi Ye, Morgan Stukes, and Anthony Boskinis.
Darah Rifkin rocked this project when she designed the BU Rock Climbing logo we have seen in the previous example.

1   Restate the project in your own words, then do some research on the industry for the company you've chosen.

2   Analyze existing logos that you see.

3   Have a brainstorming session. Identify a few good ideas.

4   Do some thumbnail sketches, work quickly, and let the ideas develop.

5   Once a sketch reveals a possible solution, go to the computer to produce it in Illustrator.
For this project, you make all the decisions for the brand. For example, determine whether the brand will be casual, mid-range, or luxurious. Identify your customers' age ranges, incomes, and so on. Then identify colors that communicate the company's products and services. Refer to Chapter 2 for a reminder of how colors convey meaning.
Check the fonts that are already in your computer: look for typefaces that suggest the personality of the brand. Remember to explore upper- and lowercase for each letter. You can expand your search to online font websites.

If developing a symbol, remember to sketch while looking at images of the subject. You can even draw over a photo that is placed and locked in Illustrator. Simplify the drawing for the best logo symbols.

6   You've already learned all the basics to produce a logo using Illustrator. This will develop your software skills as you create the mark. Step back and evaluate your work; make revisions if necessary.

7   Think about how you'd present your logo to the client or the class. Briefly explain why you chose the font, colors, and other design decisions.

8   Good luck and enjoy the process.

Tip

Follow the creative process to generate unique ideas for your logo.

Major Points Summary

— Logos convey a lot of information immediately: quality, style, value, and origin.

— Logo designs must be simple to enhance recognition, as well as ease recall and reproduction.

— Choose colors that convey meaning.

— Logos need to have a black-and-white (one color) version for versatility.

— Logotypes are defined as the name of the company or product in unique typography.

— When developing a logo, look closely at the characteristics of letters and choose a typeface that evokes the personality of the organization.

— Abstract symbols are commonly used in logos because they don't emphasize one particular area of a diverse organization.

— Abstract symbols that somehow suggest the nature of the business are called allusive symbols.

— Pictorial symbols look like the product or service they represent.

— Associative-pictorial symbols don't look like the product but have an association with the business.

— Pictorial symbols should be simplified. Choose one or two elements of a subject to emphasize in the icon (to help convey the brand dynamics).

— The creative process is a series of steps taken to successfully tackle and complete a design project. Most creative professionals develop a unique, individualized process.

— The steps suggested for beginning designers are:

1 Understand the problem
2 Get inspired
3 Brainstorm
4 Sketch
5 Produce (and revise)

Software Skills Summary

— Illustrator Advanced.

— Develop advanced skills: type options, create outlines for type, ungroup objects, lock objects, Pen tool, draw closed straight-edged shapes, Direct Selection tool, modify curved lines, Pantone swatches, Color Swatch window, eyedropper, draw closed shapes, unlock objects.

Recommended Readings

Read the informative book *Logo: The Reference Guide to Symbols and Logotypes* by Michael Evamy for a comprehensive exploration of designs from around the world.

For additional inspiration from designers who shared their work for this book, check out their websites because they may include information about their design development strategies.

Visual Themes

Stefan Sagmeister, Lou Reed poster, Tokyo, 1996

Stefan Sagmeister's designs express unique conceptual strategies and often irreverent wit. He created this poster for a Lou Reed album. The music had very personal lyrics, so he placed the text directly on the musician's face. A few years later, Sagmeister shocked the design community with a unique lecture poster. A color photo of Sagmeister's body displayed the text for the event. All the copy was cut into his torso with an X-Acto knife (an intern did the carving). This concept was meant to express the pain that accompanies some projects.

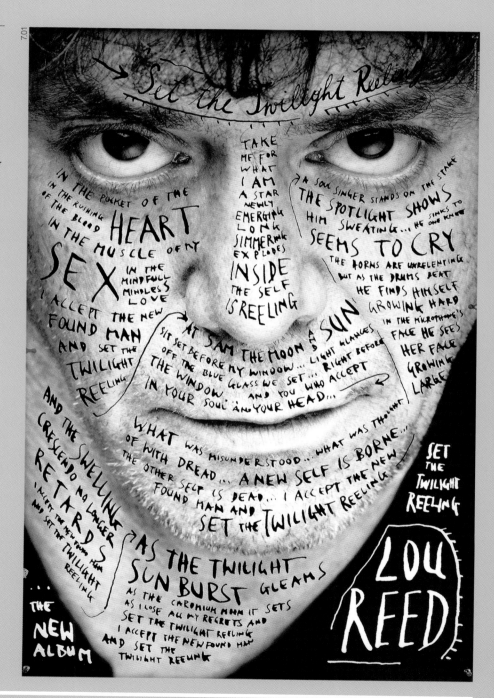

Putting it all Together

The Cochin Dingue brand is fun, bold, and consistent in every form from aprons to menus. Visual themes maintain consistency when designs occur in many different media, pages, screens, or situations. This beloved restaurant in Quebec includes signs, menus, plates, knives, aprons, and a website, and all these work as a group. Can you identify the elements that make up this brand's visual theme?

In this final chapter, you'll bring together all the design skills you've acquired to create multiple-page or -format designs. Websites, magazines, campaigns, books and packaging are examples in which

visual themes unify the overall look. Unlike a poster or advertisement, which can be understood with a glance at one page, the entire design strategy is only revealed as the reader goes through multiple pages, screens, or formats. Without visual themes—consistent use of typefaces, colors, image style, layout, and graphic elements—the audience can get confused. If confusing, designs can be off-putting and are less capable of communicating ideas.

Create a visual theme with repetition of one or two typefaces, color palette, layout, image style, and graphic elements (if used). Paper choices and printing techniques, such as die-cutting and

varnishes, can establish distinctive tactile textures. All these design elements are used to establish consistency throughout an entire design; repetition and variation of these are the keys to successful visual themes. Take another look at the Cochin collage: notice the visual themes and when they are repeated and when they are varied. For example, the backs of the business cards have either blue vertical stripes or red diagonal stripes. Cochin's visual themes have been playfully carried out on a variety of formats. The theme includes the blue, white, and red color palette, a sans-serif font, centered text on tableware, close-cropped photographs, and the graphic elements are stripes and dots.

Typeface choices are important for consistency throughout a publication or brand. Typically, one typeface is selected for the titles, another for the body text. These two are used throughout all pages and situations. Variations are introduced with contrasting size, color, style, and position.

For Bassett coffee, visual themes include a black-and-white color palette and two typefaces. An all-caps slab serif is combined with a lighter weight all-caps sans serif. The layout style on the cups and packages is a small amount of centered text surrounded by white space. The image style is consistent, the predominant graphic element is a geometric pattern, and photography is black and white when used. Notice the variations on the business cards, the pattern is interrupted, and the text is right aligned but still all caps.

Color choices establish an appropriate palette for a design. Choose enough colors for interest, but not too many, since this could dilute the overall impression. Color can be used to establish the visual hierarchy of titles and subtitles throughout the publication, adding visual interest to the page and giving the reader clues about the content.

Bright green enlivens the clever and elegant High Line logo. The symbol can be read two ways, as it both looks like an H and suggests railroad tracks. The High Line is a park created from miles of unused railroad track that run above the streets on Manhattan's west side. The floral variation for a special event (below) is appropriate because the park's seemingly wild, yet beautiful plantings are beloved.

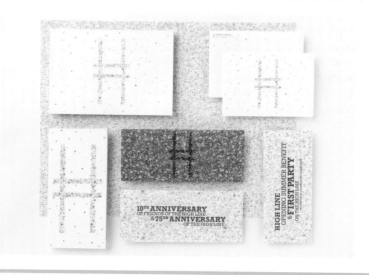

The style of images within a visual theme should be consistent. Let the image content provide the variety. If the image style is very distinctive, it is difficult to introduce another without disrupting the overall tone. Fishtank is a contemporary experimental theater where audiences experience live shows, unlike anything they've seen on stage before. These surprising visual themes play on the nouns fish and tank and use a block-print illustrative technique. Using the same style, they created a family of fish characters and use them throughout the materials for unexpected effects that are suitable for this theater.

Retro typography and the pale orange color make Mamaleh's brand memorable. This New York-style delicatessen has a line drawing of a face in the center of its logo. A low-contrast gray mosaic pattern is used as a graphic element to subtly enrich the design. The colors and fonts are so distinctive that the brand is recognized even when using a secondary font, and a different word, as you see on the Schlep tote bag. Visual themes are most effective when they are applied consistently throughout all media, from print to digital to tote bags to architecture.

The Vermont Hard Cider visual theme uses nineteenth-century design elements and typography style. These look contemporary because of the use of flat-black backgrounds and matte-gold printing. The type and graphics also work well when burned into wooden crates. Images that reference

apples and trees suggest a natural and healthy product.

A grid layout provides consistency throughout the pages of an entire publication or website. You can achieve this by establishing the width of the columns, alignment, and leading. While these particulars are repeated, there may be variation in the length or number of columns on each page. Three/six- and two/four-column grid structures are the most common because of their appropriate text-to-column width measurement. The grid establishes placement of images. Some photos fill a page, some are closely cropped to fill one unit of a grid, and other images may be silhouetted.

These pages from an American Express publication demonstrate thoughtful use of visual themes. Notice how many ways the images work together. There are similar blues in the clothing in two of the photos. The other two photos provide unity with red clothing on both figures and greenery in their backgrounds. The text layout is consistently left aligned. Narrow columns are used when next to photos, and wide text columns are used on pages without images. Notice the AmEx-brand blue is used for the subtitles and graphics.

Visual Hierarchy

Visual hierarchy helps establish a path for the reader's eye. Titles and subtitles provide the reader with summaries of content. These items are sized accordingly, as you can see in this AmEx publication. The letters of American Express are so large they don't all fit on the first spread. That is the focal point, but there are also two much smaller full names, one in the updated blue box and the other in the traditional ribbon.

Section titles are larger and often more colorful than subtitles, and both are larger than paragraph text. Page folios are even lighter in appearance. This strategy establishes visual hierarchy of the information, provides visual clues, leads the reader through the content, and enhances comprehension.

Editorial Themes

Visual themes can be purely visual or can be based conceptually on the content of the material. When based on content, these design choices are called editorial themes. Coming up with editorial themes is more challenging, but also more rewarding, for you and the audience. Look through a website or magazine and identify the visual themes. Are they related to the story or purely visual? Editorial themes are established with visual themes, but not all publications require this conceptual level of design.

The typeface for the Waltzing Matilda Centre in the Australian outback was inspired by wool bale stencils found at nearby sheep stations. This robust and unassuming font reflects the community's stalwart identity. The area's stark, rugged landscape led to the use of weathered steel for the architecture and the wayfinding system in this sunburnt country. See Chapter 1 for a photo of the museum building.

Visual themes can establish a rhythm throughout a publication or website. Designs can set an appropriate tempo—quick and lively for a stylish clothing website, or solemn and dignified for a health organization's annual report. The color, size, and placement of titles, and the length and width of the columns, contribute to the rhythm. The style of images and photo

formats (rectangles or silhouettes) affects the pace of the audience experience. Be aware that the reader's eye movement from one element to another establishes this rhythm. Inconsistency can confuse and lose the reader. Additionally, when designing with traditional media, unique papers, glosses, and embossing can convey the personality of a piece immediately. These visual and tactile characteristics make designs more memorable and therefore more likely to communicate the message.

Battersea has applied their charming and unique visual themes to a great variety of products and materials, from dog tags to posters for charity events. Notice the variation for the Muddy Dog event. They use the brand identity signature blue for the poster background color and use a muddy brown to create the dog face.

Establishing a visual identity for a fashionable beauty salon is challenging—even before one considers budget constraints—because styles will change before you're done typing the name. Gee Beauty resolved these issues by choosing a limited but elegant black-and-white palette, candid vibrant photography, and classic fonts with edgy copy. These visuals are consistent throughout, from the website to salons.

Did you recognize the eyelash curler pictured here? We started with this package design back in Chapter 1. Do you respond differently to it now? It's my hope that you come to see graphic designs in new ways—beyond initial reactions. Now you can understand underlying strategies that are used to create appealing designs. Go through this last software skill set—it will show you essential InDesign and Photoshop tricks—and you'll then be able to produce the types of designs you've seen throughout the book.

7.23

7.24

7.25

7.21

7.22

Väg is a Swedish rock band. The visual themes of their first album designs express the style of music—1970s psychedelic rock. Yellow and black amplify the style. Owned by a Stockholm design studio, the visual themes are playfully applied. This album graphic rotates when seven layers are saved as a GIF. The logo works especially well on the bass drum.

Design Analysis: Album Cover Design

Objective

Analyze album cover visual themes. What type of imagery is used? Are there photo portraits or evocative landscapes? Is the style of image consistent with social media imagery for the release of the album? Is illustration used? Is the font serif, sans serif, or novelty? How is type used in song listings? Are there any new type styles emerging on the covers of recent albums? What is the color palette? Does the design hit all the right notes?

To Do

Research an album cover or package design. Write and layout a critique in InDesign. Grab a cover image from the internet and place it in a frame on the new document. Then add a text box and type your analysis.

Stefan Sagmeister's designs often first make me laugh and then think well beyond the time I've seen them. This is a poster for a presentation and book signing of *Things I Have Learned in My Life So Far*. His studio first sent real bananas with stickers as invitations to the New York City presentation. One of the recipients, photographer Henry Leutwyler, took this photo. The studio then used it as the poster for the event when the book event was held in Colorado.

Things I Have Learned in My Life So Far includes twenty maxims. The maxims were designed using different media, installed all over

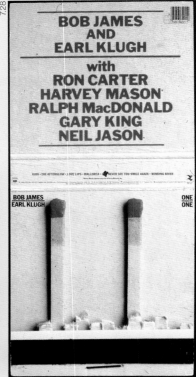

7.26

the world, photographed and published in this book. The book had fifteen different covers. One of my favorite maxims presented is "Everyone who is honest is interesting." Every seven years Sagmeister takes a year off to explore a topic. Other topics he's explored are happiness and beauty.

Paula Scher and John Paul Endress, album cover for Bob James, *H*, 1980

Paula Scher and John Paul Endress, album cover for Bob James and Earl Klugh, *One on One*, 1979

In 1974, Paula Scher began her career as an art director at CBS and Atlantic Records, making scores of groundbreaking album covers and posters. These two covers, a hotdog and a matchbook, are part of a series that features close-ups of familiar objects. Considered audacious at the time, the designs represent Scher's preference of evoking a mood or a sense of mystery and wonder, rather than showing traditional images of performers.

7.27

7.28

Design + Software Skills 7:
InDesign Advanced Images, Type, and Layout

Objectives

Establish and carry out visual themes. Use all the design strategies and skills you've gathered throughout the book to create this project. Expand your InDesign skills to multi-page designs. Design the typography in InDesign. Edit linked images.

To Do

Create an album cover and a tour poster. Place images and design text in InDesign. Establish and carry out visual themes in two formats.

Tip

In this exercise, we'll produce an album cover design and the album release tour poster. The visual themes include photographs of the Arctic. The color palette is taken from the photograph. The font is Avenir, and titles are justified, have generous tracking, and the letters are in five different colors. The variations in the visual themes include the poster image as a collage, rather than a single photo, and the locations and date text on the poster is lightweight.

1
Create a project folder

Make a new desktop folder called Project 6. Click on your desktop and select File > New Folder. Rename it Project 6.

Download the images ("ArcticWaves.jpg," "ArcticFlight.jpg," and "Portrait.png") to your desktop from http://www.bloomsbury.com/graphic-design-essentials-9781350075047. There will be three thumbnails of these images on your desktop.

Click and drag all three files over the Project 6 folder icon to put them inside. All the images are now in the folder, so the links will stay constant throughout the project.

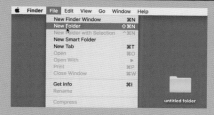

2 Create a two-page document

Open InDesign. Choose File > New > Document. Make it Print size, inches (or centimeters) and two pages. Check off Facing Pages. Leave all other default settings.

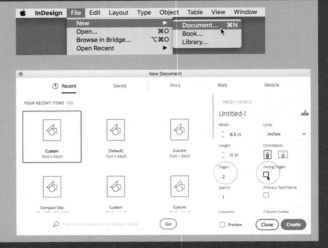

3 Draw a frame

Click on the Pages tab in the Properties panel. If the Properties panel is not showing on the right of your software window, choose Window > Workspace > Reset Essentials.

Select the Frame tool and click on the file to open the Frame panel. Make the cover dimensions 4.72 inches × 4.72 inches (11.98 cm × 11.98 cm).

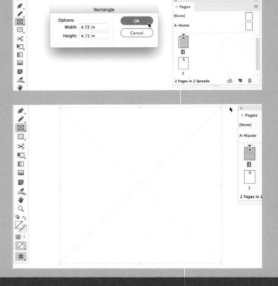

4 Place the image

Click on the frame. Choose File > Place and select the "ArcticWaves.jpg."

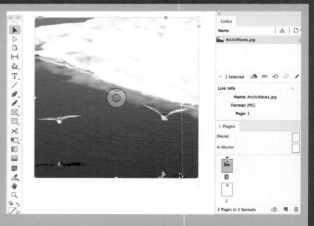

5 Resize the image

Resize the photo proportionately to fit the cover format. Choose Object > Fitting > Fill Frame Proportionately.

Tip

InDesign displays images at low resolution for faster processing times. Your high-quality photos and vector images may appear rasterized but will print correctly. To view images in high resolution, click on the image and choose Object > Display Performance > High Quality Display.

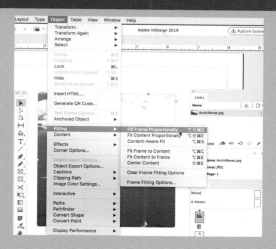

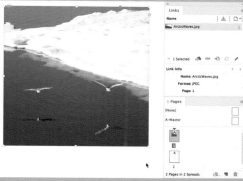

6 Add the titles

Make a text box below the left shadow. Place your Type tool cursor just below and to the left; end the box about mid-wing to the right.

Using all capitals, type "ARCTICLOUDZ" on the first line, press Return and type "ARCTICWAVES."

Note the unique spelling in the top line: one C and a Z, instead of an S; also note the lack of word spacing on both lines.

In the Properties panel, make the font Futura Medium Condensed, 12 pts. In the lower right box, make the Tracking 100. Under Paragraph, select the icon for Justify with the last line center aligned.

7 Apply colors

At this point, we will create a color palette for the titles from the photo.

Select the Color Theme tool and click on the image. The software will create a five-color palette, providing a range of hues from the photo. Now, click on the Swatch icon (to the right of the suggested colors) to add this palette to your file's color swatches.

Open the color palette to apply the colors to the letters. Choose Window > Color > Swatches to see the swatches. Find the Colorful Theme folder and click to open it and apply the colors.

Select the A in the title first, click on the lightest color in the Colorful Theme. Then select the R and click on the next darker value. Continue through the letters of the name and title. Make both As and the W the lightest color for optimum legibility.

Tip

Nice work! You can view your design without frames and guides by pressing W. To see the frames and guides, press W again.

Like Illustrator, InDesign has the Change Case menu. Choose Type > Change Case to view variations when designing your projects; make the most effective case choice for your font.

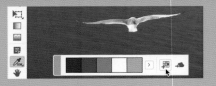

8 Make the poster

Click on the second page icon in the Pages panel. Select the Frame tool. Click and make a frame using the international poster standard format: 6.5 inches × 10 inches (16.52 cm × 25.4 cm).

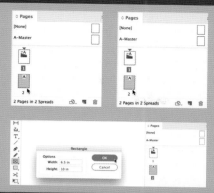

9 Place the image

Click on the frame. Choose File > Place and select "ArcticFlight.jpg."

This collage was made to fit the poster format.

Tip

When using an image that is larger than your frame, resize by choosing Object > Fitting > Fit Content Proportionately.

Confirm you have a high-resolution image by selecting Object > Image > High Quality Display.

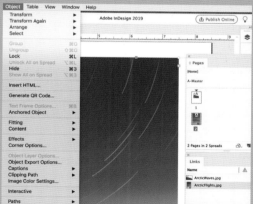

10 Lock the image

Click on the image and lock it into position by choosing Object > Lock (or press Command + L). Locking keeps the collage in place as you add other elements to the design. You can always unlock by choosing Object > Unlock All on Spread.

11 Add a round frame

Under the Frame tool, choose the Ellipse Frame tool. Hold your Shift key down, click on the image, and drag to draw a circle. This circle is 1.12 inches (2.84 cm) in diameter; make your circle approximately the same size. Position it towards the top left of the page, centered above the bird and jet as you see here.

12 Ellipse frame, place portrait, and resize

Click on the Ellipse Frame. Choose File > Place and select "Portrait.png" from your Project 6 folder. Images appear in frames at actual size. Resize the photo by choosing Object > Fitting > Fit Content Proportionately. If necessary, adjust the portrait's position in the frame by using the white Direct Selection tool.

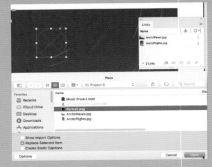

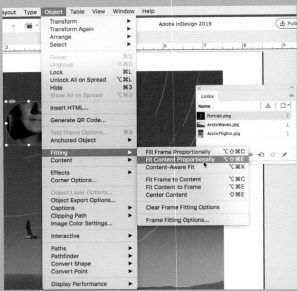

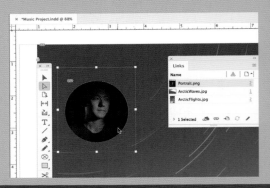

13 Add the titles

First copy the title text box from page one onto the page two. Click on the page one icon, select the text box, press Command + C to copy.

Now double-click on the second page icon in the Pages panel. Press Command + V to paste the title text box onto page two.

Select the titles with the Type tool, change the font size to 20 pts in the Properties panel. If the panel is not showing, choose Windows > Workspace > Reset Essentials.

Add the tour text. Using the Type tool, click below the left edge of the title box. Type "2021 WORLDWIDE TOUR." Make the font size 12 pts.

Adjust the tracking on "2021" and "Tour," so they extend to align with the titles. This tracking is set to 120; yours may need slight adjustments.

Change the colors of these letters as we did in step 7. As you see here, "2021" and "Tour" are the lightest color, and the "Worldwide" letters are the medium blues.

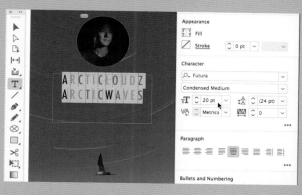

14 Add the cities and dates

Use the Type tool to drag a text box on the right middle of the page. The box will be about 2 inches wide × 5 inches tall (5 cm × 12 cm) and end above the gulls. You can add these cities or make your own list of favorites. The text is Futura PT, Light, 11 pts, Auto Leading (13.2). This text is the lightest color from the color theme in the swatches.

Next, make a narrow text box for the dates to the left. This box is 0.5 inches wide (1.27 cm). Add the dates. This text is the lightest color from the color theme in the swatches.

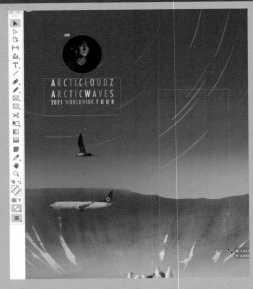

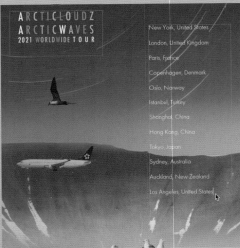

15 Align the columns

Place the two columns in alignment using Smart Guides. Confirm the baselines are aligned by dragging a guide at the top of "ARCTICWAVES" and the top date and city boxes.

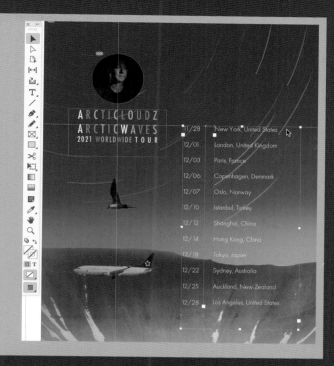

16 Add the website address

Add the web address to the bottom of the poster. For optimal visual hierarchy, the text is now all lowercase Futura Light, 10 pts. Notice the change from dark text to light text due to the ice color behind it.

Tip

All your placed images are listed in the Links window. If you rename a file or move it, InDesign will prompt you to fix the link. Open the Links panel under Window > Links. Relink by clicking on the file name, then clicking on the Relink icon in the bottom row of the panel.

17 Finishing off

Save your work as an InDesign .indd format. Also Export as a .pdf. Choose File > Export > pdf.

Great Work! Congratulations on completing our final, complex design and software skill set.

This cover and poster design were created by student and musician Ziqian Liu. The design, and the music, were inspired by his travels to the Arctic Circle when he was a student at Boston University. These are his photos, and you can hear his tracks at soundcloud.com/arcticloudz. Enjoy.

Design Project 6:
Album Cover and Tour Poster Designs

Design an album cover and a related tour or concert poster. The music on the album may be a redesign of an existing album, a mix, a festival, your own, or a friend's music.

You will establish visual themes and visual hierarchy in this project. The project provides an opportunity to lay out similar content into two different formats: square (cover) and rectangular (poster).

To establish the visual themes, choose one or two typefaces—one font may suggest the tone of the music and another for the supporting copy, including the tour dates, cities, and venues. Or, use your favorite professional font and create visual hierarchy with different sizes, weights, and color. Find images that work together; choose a color palette from the images and/or colors that help express the style of music.

Copy Specifications:
Cover: Band name and album title.
Poster: Band name, album title, the word "tour" (or "'concert" if only one date), cities, dates, and the band web address.

Tip
Follow the creative process:
1. Research and analyze album cover designs.
2. Brainstorm about your music selection.
3. Listen to the music.
4. Research images and sketch your concepts for each format, the cover, and poster.
5. Use Photoshop and Illustrator (as needed), and place images into InDesign to produce the project. Add the text and design for good visual hierarchy of information.

Student Album Covers and Tour Posters

For his final project, Ziqian Liu used his own photographs from a trip to Svalbard. An archipelago between mainland Norway and the North Pole, Svalbard is in one of the northernmost regions on earth. The area inspired Ziqian's music composition, *Arctic Waves*, an EP with five individual pieces that correspond to the five photographs he used in the cover and the poster collage. Ziqian explains, "The birds and jets are particularly relevant in demonstrating a sense of freedom and the concept of voyage, which relates to the idea of touring around the world." Regarding his design strategies, he says, "The album cover and the tour poster manifest the effectiveness of asymmetrical

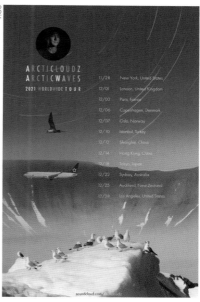

design." He used Futura Medium for titles and a lightweight version for the cities and dates. Ziqian selected colors for the text from the photos. In the skill set, we used the Color Theme tool to provide a color palette from the cover photo.

This cover and poster, by student Nicole Layne, was inspired the artist Tash Sultana's tattoo. Tash is a singer-song writer as well as a multi-instrumentalist. All her EP album covers have been completely hand-drawn. Nicole found the imagery was demonstrative of the type of music Tash produces. Nicole's design concept was to reimagine the artist herself in an illustration.

Nicole used Illustrator to hand draw the vector images shown in both the album and the poster. For the album, she used the pen tool to trace a photo of the musician showing her elephant tattoo. To continue this visual theme, she enlarged the scale of the tattoo and recreated the elephant using the pen tool; then she added an accent color for the poster. The font is Myriad Pro, and Nicole added horizontal lines to customize Tash's name.

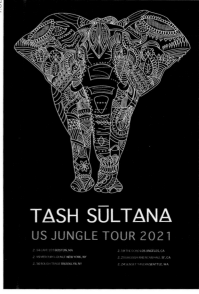

Lindsay Kerr created a cover and poster design for the Coldplay song, "The Scientist." The song is about the bad effects we have on the environment. In particular, Lindsay was thinking about the lyric, "Take me back to the start," a time before environmental issues were a problem. Lindsay wanted her design to raise awareness of how pollution directly impacts us, providing immediacy to the global threat.

Her concept for the design was to use ink that reacts to the amount of pollution in the air. Those who bought the album or see the poster outside in their neighborhoods would see how much pollution is in the area. The darker the cityscape inside the lungs appears, the more pollution in the air and, consequently, in our lungs and

bodies. The image of the lungs is meant to warn people that pollution affects our bodies. Lindsay found the photo on UnSplash.com and drew the illustrations using software on her tablet. She chose Balboa, a display sans-serif font, for her text.

Graduate Teaching Assistant Yuling Lu, who was tremendously helpful preparing this book, created this design when she was a student. For her project, Yuling redesigned a band's album artwork. The original cover had a graphic of a woman's profile, which Yuling reoriented onto the horizontal axis. Transposed, the image reads as sound waves. The profile of a face becomes less immediate, but once noticed, it adds depth and interest to the album cover. Yuling hopes viewers will "be surprised and feel that the design is worth spending time with, and thus become more interested in the music itself." By using the font Impact and all capitals in the band name, the *A* and *W* seem to lean against each other, while

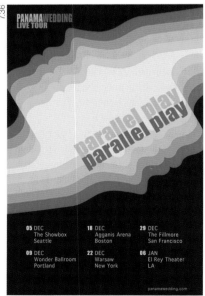

lowercase letters in the album title, *Parallel Play*, enhance the mirrored effect. Panama Wedding is a retro Synthpop band, and the redesign's color palette expresses a fitting retro vibe.

Major Points Summary

— Visual themes unify the overall look of multiple-page publications such as apps, websites, magazines, advertising campaigns, annual reports, and packaging.

— Establish visual themes with a specific color palette, typefaces, image style, layout, and graphic elements.

— Inconsistent visual themes confuse and potentially lose the audience.

— Typically, one or two typefaces are used throughout a design; variations are introduced with type family styles, color, size, texture, and placement.

— Choose colors that are appropriate for your client but not too many because it can weaken the overall cohesiveness.

— The style of images should have consistency throughout a design.

— The grid provides consistency throughout a multiple-page design with width of columns, alignment format, leading, and position of images.

— Visual hierarchy establishes the relative importance of the copy and leads readers' eyes through the text.

— Editorial themes are visual themes based on the content of the material.

— Unique papers, inks, glosses, and embossing can convey the personality of the piece immediately.

— Tactile effects make for more engaging and, therefore, more effective communication, but they can also be more expensive to produce.

— Visual and editorial themes can be used throughout media campaigns and should be consistent across all media from print to websites to architecture.

Software Skills Summary

InDesign advanced skills: specialized document format; pages window; square, circular, and rectangular frames; place images; resizing; fill frame proportionately; fit content proportionately; display performance; format text; using the eyedropper; change case; color theme tool; swatches; transform and move image in the frame with the Direct Selection tool; draw guides, draw perfect circles from the center; copy and scale formatted titles; tracking; Links panel; relink; print and export .pdf.

Recommended Readings

TED Talk Stefan Sagmeister, "The power of time off."

https://www.ted.com/talks/stefan_sagmeister_the_power_of_time_off?language=en

Things I Have Learned in My Life so Far, updated edition, by Stefan Sagmeister, Daniel Nettle, Nancy Spector, Steven Heller. This revised and updated edition includes all the aphorisms from the first book, along with an additional forty-eight pages of new ones and recent exhibitions.

Paula Scher: Works was co-edited by Tony Brook and Adrian Shaughnessy. The book organizes her design work chronologically into several thematic sections. Get it at the library!

Appendix

Keyboard Shortcuts: Adobe InDesign, Illustrator and Photoshop

Function	Mac Keystrokes	PC Keystrokes
close a file	⌘ + W	ctrl + W
copy	⌘ + C	ctrl + C
cut	⌘ + X	ctrl + X
open a new file	⌘ + N	ctrl + N
open a file	⌘ + O	ctrl + O
paste	⌘ + V	ctrl + V
print	⌘ + P	ctrl + P
save	⌘ + S	ctrl + S
select all	⌘ + A	ctrl + A
show rulers	⌘ + R	ctrl + R
swap foreground/background	X	X
switch to Type tool	T	T
undo	⌘ + Z	ctrl + Z
view file full size	⌘ + 0 (zero)	ctrl + 0 (zero)
zoom in	⌘ + +	ctrl + +
zoom out	⌘ + -	ctrl + -

Keyboard Shortcuts: Adobe InDesign

Function	Mac Keystrokes	PC Keystrokes
align left, right, or center	⌘ + shift + L or R or C	ctrl + shift + L or R or C
all caps	⌘ + shift + K	ctrl + shift + K
deselect all	⌘ + shift + A	ctrl + shift + A
find/change	⌘ + F	ctrl + F
hide frames	⌘ + ctrl + H	ctrl + H
hide guides and margins	⌘ + ;	ctrl + ;
group objects	⌘ + G	ctrl + G
ungroup objects	⌘ + shift + G	ctrl + shift + G
lock object	⌘ + L	ctrl + L
unlock object	⌘ + opt + L	ctrl + shift + L
place	⌘ + D	shift + D
spell check	⌘ + I	shift + I
switch to Direct Selection	A	A
switch to Selection tool	V	V
switch to/from preview	W	W
temporary switch to Selection	⌘ (press and hold)	ctrl (press and hold)

Keyboard Shortcuts: Adobe Illustrator

Function	Mac Keystrokes	PC Keystrokes
align left, right, or center	⌘ + shift + L or R or C	ctrl + shift + L or R or C
deselect all	⌘ + shift + A	ctrl + shift + A
deselect Snap to Point	⌘ + opt + '	ctrl + alt + '
group objects	⌘ + G	ctrl + G
ungroup objects	⌘ + shift + G	ctrl + shift + G
hide guides	⌘ + ;	ctrl + ;
lock object	⌘ + 2	ctrl + 2
unlock object	⌘ + opt + 2	ctrl + alt + 2
place	⌘ + shift + P	ctrl + shift + P
spell check	⌘ + I	ctrl + I
switch to Selection tool	V	V
switch to Direct Selection	A	A
temporary switch to Selection	⌘ (press and hold)	ctrl (press and hold)

Keyboard Shortcuts: Adobe Photoshop

Function	Mac Keystrokes	PC Keystrokes
cycle through open files	ctrl + tab	ctrl + tab (check PC)
deselect	⌘ + D	ctrl + D
exit cropping	Crop tool + esc	Crop tool + esc
free Transform	⌘ + T	ctrl + T
group layers	⌘ + G	ctrl + G
ungroup layers	⌘ + shift + G	ctrl + shift + G
magnify 100%	Double-click Zoom tool	Double-click Zoom tool (PC?)
multiple undos	⌘ + opt + Z	ctrl + alt + Z
open the Image Curves window	⌘ + M	ctrl + M
open the Image Size window	⌘ + opt + I	ctrl + alt + I
temporary switch to Move tool	⌘ (press and hold)	ctrl (press and hold)

Glossary

abstract symbol—in logos, a nonrepresentational shape.

Acrobat—Adobe software for creating and viewing PDF documents.

additive color—the color system by which we view colors projected by light (digital displays, TVs). The computer creates color by illuminating red, green, and blue (RGB) phosphors; when all colors are combined (added) they create white.

Adobe—software company that produces the Creative Cloud software, which includes Photoshop, Illustrator, InDesign, Bridge, and Acrobat. Additional software includes Premier for video editing.

allusive symbol—in logos, a nonpictorial shape that alludes to the organization it identifies.

associative symbol—in logos, icons that depict an object that is related to the organization it represents.

asymmetrical type/format—a layout in which each line is placed in a varying position.

bitmap—the grid of colored squares that composes a digital image. The grid is the map; each bit is a color. JPEG, PNG, and GIF files are bitmap images.

brainstorm—a session in which a group of people generates ideas about a subject, without criticism, then afterwards evaluate and pare down the results. Used to come up with fresh ideas.

brightness—the amount of white in a color.

centered type/format—a layout in which all the lines align to the center. A formal alignment, it is considered difficult to read more than several lines in this format.

CMYK—cyan, magenta, yellow, black. Called the four-color printing process, these inks are combined to produce a full spectrum of colors. Printed images are prepared in CMYK mode.

Cold-press paper—paper with a somewhat rough surface.

comp—short for comprehensive or composition layout. The layout of a proposed design; or a stock photograph prior to rights agreement and purchase.

compression—reduces the size of an image file by eliminating data.

contour type/format—layouts that follow a shape or image.

contrast—pronounced differences of elements in a design. It is used to attract attention, create focal points, add visual interest, and indicate hierarchy of information.

divine proportion—in art, it is considered to be the most aesthetically pleasing rectangular proportion; Greeks based the Parthenon on this. Also called the golden ratio.

dpi (dots per inch)—measurement of digital images. One dot equals one pixel.

editorial theme—visual themes that are conceptually consistent with the subject in the design.

elements of design—the most basic components of a design: color, direction, line, shape, size, texture, value.

EPS (Encapsulated Post Script)—uncompressed high-quality image file format.

Flash—Adobe animation software.

focal point—the area of the page that initially attracts attention.

font—the complete set of characters in any given size and style. Also refers to the software version of typefaces.

frame—in InDesign, frames provide placeholders and formats for images and text.

full-bleed—ink runs beyond the margins of the design so that, when the paper is trimmed, the image goes to all four edges.

gamut—the range of colors in a color space.

GIF (Graphics Interchange Format)—compressed file format. Use for logos, text, and simple graphics for screen display only. GIF compression eliminates variations of colors to reduce the file size. GIF files are bitmap (or raster) images.

glyphs—special typographic characters; can be shapes, unusual letters, or punctuation.

golden ratio—an irrational number, approximately 1.618. In art, it is considered to be the most aesthetically pleasing rectangular proportion. Greeks based the Parthenon on this number, also called the divine proportion.

golden rectangle—the typical form of the golden ratio or divine proportion in art and architecture.

grid layout—all the elements on a page, or throughout a document, are aligned to an underlying skeletal structure. The grid is drawn with horizontal and vertical lines.

hanging punctuation—adjusting the position of quotation marks, periods, and other punctuation so that the letters are aligned and the punctuation "hangs" beyond the sharp edge of the lines of type.

hi-res—short for high-resolution file quality, used for digital images that will be printed.

hot-press paper—paper with a smooth surface.

hue—another word for color, it also refers to the name of a color.

Illustrator—Adobe software for creating single- or multi-page vector images.

image banks—commercial resources for images including photography, illustration, and video.

InDesign—Adobe multiple-page layout and prepress software.

JPEG (Joint Photographic Experts Group)—the most common compression file format used for photographs and complex illustrations intended for screen displays. File compression results in smaller file sizes. JPEG files can be saved with low, medium, and high quality.

justified type/format—a layout in which all the lines align on the left and right margins. Considered the most legible format for reading large amounts of text.

kerning—adjusting the area between uppercase letters for ideal spacing in titles and logos.

layers—used to keep file contents separate. In Photoshop, they allow non-destructive edits and effects to various components of a design; they keep typography from being affected by changes to images. In Illustrator and InDesign, layers are used to organize the content of complex files. Layers can be transformed, grouped, made invisible, and exported into Acrobat.

leading—pronounced "led-ing." The vertical space between lines.

left-aligned type/format—a layout in which all the lines align on the left margin. Sometimes called flush left/ragged right.

logotype—the name of an organization or product in unique typography.

low-res—short for low-resolution file quality, used for screen display of digital images.

mahalo—"thank you" in Hawaiian.

Modern typefaces—characterized by a distinct contrast between thick and thin strokes and hairline serifs set at abrupt right angles to the stems.

mono-weight—a typeface style in which there is no variation in the width of the strokes and stems of letterforms. Most sans-serif typefaces are mono-weight, leading to better legibility for onscreen text.

novelty typefaces—unique fonts rich in personality, they provide a distinctive voice to a design.

Old Style typefaces—characterized by gradual blending to slanted serifs and low contrast between thick and thin strokes.

Pantone—the leading international color company. It publishes the standard color matching system (PMS), which guarantees consistency of colors. Also called spot color.

path layout—all the elements on the page are arranged to lead the eye through the layout in a meaningful way.

PDF (Portable Document Format File)—a format that displays all the elements of a design without requiring the software that was used to create it. Can be used to transfer designs via the internet, for large multiple-page document access through websites, and for printing. Can be saved in low to high resolutions.

Photoshop—the industry-standard software for manipulating images by Adobe.

pictorial symbol—in logos, an icon that looks like the product or service of the organization it represents.

pixel—one tiny square of a raster (or bitmap) image, displays one color. It is the unit of measurement for onscreen digital images.

PMS—Pantone Matching System.

PNG (Portable Networks Graphics)—a compression file format that works well when a transparent background is required around an image.

ppi (pixels per inch)—measurement of digital displays.

raster image—an image made of pixels, often created or edited in Photoshop.

rasterization—the process of converting a vector image to an image made of pixels.

rebus—a visual puzzle made of a string of letters and pictures that represent sounds in place of words or syllables.

resolution—measured by the number of pixels per inch. High-resolution refers to sharper, more detailed images. Low resolution is typical on computer displays because it requires smaller file sizes and faster processing times.

RGB—red, green, blue; the primary colors of the additive system. Onscreen images appear in RGB mode.

right-aligned type/format—a layout in which all the lines align on the right margin. Considered difficult to read over more than a few lines.

rights-managed images—images whose cost is calculated based on the intended size, placement, duration, and geographic distribution of use.

royalty-free images—a collection of images that may be used repeatedly in return for a one-time fee.

sans serif—a typeface of letters without serifs. Also refers to the category of typefaces that do not have serifs. Sans means "without" in French.

saturation—the intensity of the color, it indicates the amount of gray in a color.

screen shot—a digital image of a computer screen display.

scripts—typefaces that appear to be hand-drawn.

semiotics—the study of signs and symbols and how they are used and interpreted.

serif—the notches at the end of letterforms. Also refers to this category of typefaces.

Slab Serif—typeface with serifs that are the same weight as the vertical and horizontal strokes. Also refers to the category of Slab Serif typefaces.

spot color—a particular ink color, rather than a combination of CMYK.

subtractive color—the pigment-based color system. Ink absorbs (subtracts) a portion of the color spectrum, and what is not absorbed reflects back to the eye, resulting in the colors we see. When the primary colors, red, blue, and yellow, are all absorbed by a pigment the result is black.

symmetrical layout—all the elements on the page align along the center axis.

tactile—a texture than can be felt.

thread—in InDesign, this function allows text to flow throughout multiple frames or columns.

thumbnail sketch—several small, quick sketches used to develop and evaluate layouts.

TIFF (Tagged Image File Format)—high-quality, high-resolution, uncompressed image file format. Typically used for printing. All the images in this book are TIFF format.

transitional typefaces—characterized by deeper bracketing to the serifs and more contrast than Old Style fonts. Considered to be highly legible.

type family—a set of fonts that contains all the variations of a particular typeface.

unjustified type format—text that is not aligned on both margins. It can be flush left, right, centered, or asymmetrical.

value—the relative lightness or darkness of a color.

vector graphics—graphics produced in Illustrator and InDesign that can be resized without losing quality; they are made of points, lines, and curves using mathematical equations (rather than pixels).

visual hierarchy—the interpretation of the relative importance of the words on the page with regard to the size, color, and position of the text.

visual theme—the repetition and variation of typefaces, colors, layouts, and image styles to create a cohesive design throughout multiple-page publications.

x-height—the height of a lowercase letter.

Credits

1.01: Lucien Bernhard, photo by © Historical Picture Archive/CORBIS/Corbis via Getty Images

1.02: Gee Beauty; GJP Advertising – Lisa Greenberg, Trevor Schoenfeld

1.03: Ink Creative Director: Matthew Squadrito, Terry Squadrito

1.04: Courtesy of Pentagram Design, Creative Director: Paula Scher/Pentagram

1.05: Courtesy of Pentagram Design, Creative Director: Abbott Miller/Pentagram

1.06: TM & © 2006 Survivor Production, LLC., Used with the kind permission of Survivor Productions, LLC. Survivor Production, LLC.

1.07: TM & © 2006 Survivor Production, LLC., Used with the kind permission of Survivor Productions, LLC. Survivor Production, LLC.

1.08: @SNASK

1.09: Courtesy of Pentagram Design, Creative Director: Paula Scher/Pentagram

1.10: Emiliano Ponzi

1.11: Photos: Michael Piazza, Katie Noble. Layout: Ilene Bezahler. Styling: Catrine Kelty, Sarah Blackburn

1.12 and 1.13: Pentagram Design

1.14: *The Other Americans*, by Laila Lalami, published by Bloomsbury Circus, 2019.

1.15: Jennifer Lucey-Brzoza

1.16: Courtesy of Pentagram Design, Creative Director: Paula Scher/Pentagram

1.17: Robin Rotman for Desk Plants, DeskPlants.com. 2018

1.18: Agency: LMG Audace & créativité. Creative Direction: Jacques-Dominique Landry, Louis Martin; Art Direction: Mathieu Tremblay; Design: Mathieu Tremblay, Laurent Grislain; Photography: Laurent Grislain; Client: Restos Plaisirs Group

1.19: ©dotdash ©Florain Groehn Photography

1.20: ©dotdash ©Florain Groehn Photography

1.21: Courtesy of Pentagram Design, Creative Director: Eddie Opara/Pentagram

1.24: Joshua Sweeney, Shoot for Details

2.01: Copyright Michael Schwab Studio

2.02: Emiliano Ponzi

2.03: Design: Stefan Sagmeister

2.04: Courtesy of Pentagram Design, Creative Director: Paula Scher/Pentagram

2.05: Design Ranch

2.06: Design Ranch

2.07: Studio Feixen

2.08: Design Ranch

2.09: *High Heel*, by Summer Brennan, published by Bloomsbury Academic, 2019. Part of the Object Lessons series.

2.10: Emiliano Ponzi

2.11: Studio Feixen

2.12: Copyright©2016 Mortise Design.LLC

2.13: Forward Festival 2017 – Timetable Vienna. Copyright Zwupp

2.14: @SNASK

2.15: Kao Wei Che

2.16: Design Ranch

2.17: ©dotdash ©Florain Groehn Photography

2.18: Kao Wei Che

2.19: Firm: RK Venture. Client: Horsemen's Feed & Supply Co.; Creative Director/Designer/Copywriter: Rudi Backart

2.20: Courtesy of Pentagram Design, Creative Director: Emily Oberman/Pentagram

2.21: Courtesy of Pentagram Design, Creative Director: Paula Scher/Pentagram

2.22: Courtesy of Pentagram Design, Creative Director: Abbott Miller/Pentagram

2.23: House Industries

2.24: Design Ranch

2.25: Courtesy of Pentagram Design, Creative Director: Eddie Opara/Pentagram

2.26: Creative Director: Matthew Squadrito, Terry Squadrito

2.28 and 2.29: Creative Direction: Stefan Sagmeister; Art Direction: Iannis Kandyliaris; Design: Wade Jeffree; Edition: Bjarke Ingels; Book & Exhibition Concept: Bjarke Ingels, Iannis Kandyliaris, Alana Goldweit & Kai-Uwe Bergmann

2.30: Courtesy of Pentagram Design Creative Director: Michael Bierut/Pentagram

2.31: *Lateral Cooking*, by Niki Segnit, published by Bloomsbury Publishing, 2018. Design and illustration by A Practice for Everyday Life.

2.32: Executive Creative Director: Jorge Brunet-García; Creative Director: Eduardo Sarmiento; Art Director: Bianca Borghi

2.33: Photos: Michael Piazza, Katie Noble. Layout: Ilene Bezahler. Styling: Catrine Kelty, Sarah Blackburn

2.34: Jennifer Lucey-Brzoza

2.35: Photography © Natural Highs Festival, Design: Serafim Mendes

2.36: Courtesy of Pentagram Design, Creative Director: Emily Oberman/Pentagram

3.01: Courtesy of Pentagram Design, Creative Director: Paula Scher/Pentagram

3.02: @SNASK

3.12: House Industries

3.13: House Industries

3.15: Piet Zwart, © DACS 2020

3.16: Courtesy of Pentagram Design, Creative Director: Abbott Miller/Pentagram

3.18: Courtesy of Pentagram Design Creative Director: Michael Bierut/Pentagram

3.19: Design Ranch

3.20: Photos: Michael Piazza, Katie Noble. Layout: Ilene Bezahler. Styling: Catrine Kelty, Sarah Blackburn

3.21: Design Ranch

3.22: ©dotdash ©Florain Groehn Photography

3.23: Emiliano Ponzi

3.24: ©dotdash ©Florain Groehn Photography

3.25: Courtesy of Pentagram Design, Creative Director: Abbott Miller/Pentagram

3.26: Courtesy of Pentagram Design, Creative Director: Michael Bierut/Pentagram

3.27: Pentagram Design

3.28: Design Ranch

3.29: Executive Creative Director: Jorge Brunet-García; Creative Director: Eduardo Sarmiento; Art Director: Bianca Borghi

3.30: Studio Feixen

3.31: Photos: Michael Piazza, Katie Noble. Layout: Ilene Bezahler. Styling: Catrine Kelty, Sarah Blackburn

3.36: Tao Xuzhi

3.37: Yuling Lu

3.38: Jessica Hudson

3.39: Wasin Kittiwan

3.40: Hayley Levesque

3.41: Pieter Melotte

3.43: Raisa Acloque

3.45: Tao Xuzhi

3.44: Raymond Yu

4.01: Milton Glaser, photo by Michael Ochs Archives/ Getty Images

4.02: Art Direction: Underline Studio; Photography: Scott Conarroe; Author: Robert Bean; Publisher: Prefix Institute of Contemporary Art; Photo-documentation: Underline Studio © Prefix Institute of Contemporary Art, 2012

4.03: Emiliano Ponzi

4.04: *Lateral Cooking*, by Niki Segnit, published by Bloomsbury Publishing, 2018. Design and illustration by A Practice for Everyday Life.

4.05: Copyright©2016 Mortise Design.LLC

4.06: Courtesy of Pentagram Design, Creative Director: Natasha Jen/Pentagram

4.07: Emiliano Ponzi

4.08: © Kasnot Medical Illustration. All rights reserved.

4.09: Design Ranch

4.10: Pentagram Design

4.11: Copyright © 2016 Mortise Design.LLC

4.12: Executive Creative Director: Jorge Brunet-García; Creative Director: Eduardo Sarmiento; Art Director: Bianca Borghi

4.13: Yuling Lu

4.14: Herbert Matter, photo by Found Image Holdings/Corbis via Getty Images

4.15: Creative Director: Matthew Squadrito, Terry Squadrito

4.16: Franziska Barczyk

4.17: Miguel Porlan

4.18: Executive Creative Director: Jorge Brunet-García; Creative Director: Eduardo Sarmiento; Art Director: Bianca Borghi

4.19 and 4.20: Jason Zucco

4.22 and 4.23: Emily Hanks, www.homebehindtheworldahead. com

4.24: Photos: Michael Piazza, Katie Noble. Layout: Ilene Bezahler. Styling: Catrine Kelty, Sarah Blackburn

4.25: Studio Feixen

4.26, 4.27 and 4.28: Firm: RK Venture; Client: Horsemen's; Feed & Supply Co.; Creative Director/Designer/ Copywriter: Rudi Backart

4.29: Sprout Studios Boston

4.30 and 4.31: © dotdash © Florain Groehn Photography

4.32, 4.33 and 4.34: Copyright © 2016 Mortise Design.llc

4.35: House Industries

4.36: Photos: Michael Piazza, Katie Noble. Layout: Ilene Bezahler. Styling: Catrine Kelty, Sarah Blackburn

4.37: Design Ranch

4.38: Courtesy of Pentagram Design, Creative Director: Emily Oberman/Pentagram
Pages 118-121: Getty Images, Digital Vision
Pages 122-126: Jason Zucco
Pages 127-132: Jason Zucco

5.01: Photo by Movie Poster Image Art/Getty Images

5.02: Emiliano Ponzi

5.03: Copyright©2016 Mortise Design.LLC

5.04: Courtesy of Pentagram Design, Creative Director: Paula Scher/Pentagram

5.05: Creative Director: Matthew Squadrito, Terry Squadrito

5.06: Courtesy of Pentagram Design Creative Director: Michael Bierut/Pentagram

5.07: Courtesy of Pentagram Design, Creative Director: Paula Scher/Pentagram

5.08: Emiliano Ponzi

5.09: Photos: Michael Piazza, Katie Noble. Layout: Ilene Bezahler. Styling: Catrine Kelty, Sarah Blackburn

Pages 41-145: Kori Mausner, Design, Mitch Weiss, Photography

5.11: Tatte website, © 2020 Tatte Bakery And Cafe. All rights reserved. Site by Amanda Jane Jones & Gemma Haylett Web Development

5.12: *You Will Be Safe Here*, by Damian Barr, published by Bloomsbury Publishing, 2019.

5.13: Jamie Balkin

5.14: Samantha Lee

5.15: Courtesy of Pentagram Design, Creative Director: Eddie Opara/Pentagram

5.16: Courtesy of Pentagram Design, Creative Director: Paula Scher/Pentagram

5.17: @SNASK

5.18, 5.19 and 5.20: Robin Rotman for Desk Plants, DeskPlants.com. 2018.

5.21, 5.22 and 5.23: Design Ranch

5.24 and 5.25: Photos: Michael Piazza, Katie Noble. Layout: Ilene Bezahler. Styling: Catrine Kelty, Sarah Blackburn

5.26: Photos: Michael Piazza, Katie Noble. Layout: Ilene Bezahler. Styling: Catrine Kelty, Sarah Blackburn

5.27: Poster, der Film [Film], 1959-60. Designed by Josef Digitale (1) (A) Muller-Brockmann, (Swiss, 1914-1996). Offset lithograph on paper, 50 3/8 × 35 11/16 in. (1280 × 906 mm). Museum purchase from General Acquisitions Endowment Fund, 1999-46-2. Photo: Matt Flynn © Smithsonian Institution. New York, Cooper-Hewitt – Smithsonian Design Museum. © 2019. Cooper-Hewitt, Smithsonian Design Museum/Art Resource, NY/Scala, Florence

6.01: Courtesy of International Business Machines Corporation, © 1981 International Business Machines Corporation.

6.02: Sprout Studios Boston

6.03: BMW Group

6.04, 6.05, 6.06, 6.07 and 6.08: Courtesy of Pentagram Design, Creative Director: Abbott Miller/ Pentagram

6.09: Image Courtesy of Chermayeff & Geismar & Haviv

6.10 and 6.11: Jennifer Lucey-Brzoza

6.12: @SNASK

6.13: Design Ranch

6.14 and 6.15: Courtesy of Pentagram Design, Creative Director: Eddie Opara/Pentagram

6.16: Studio Feixen

6.17: Image Courtesy of Chermayeff & Geismar & Haviv

6.18: Design Ranch

6.19 and 6.20: Jennifer Lucey-Brzoza

6.21: Image Courtesy of Chermayeff & Geismar & Haviv

6.22: Image Courtesy of Chermayeff & Geismar & Haviv

6.23: Copyright©2016 Mortise Design.LLC

6.24: Mike & Patty's, LLC, Studio: Aurora Studio

Color Connotations

6.25: Tank Design Inc. Cambridge MA – Principal Designer Brandon Miller

6.26 and 6.27: Pentagram Design

6.28 and 6.29: Courtesy of Pentagram Design, Creative Director: Paula Scher/Pentagram

6.30, 6.31 and 6.32: Jennifer Lucey-Brzoza

6.33: Design Ranch

6.34, 6.35, 6.36 and 6.37: Creative/Art Direction: Jessica Walsh & Stefan Sagmeister; Animation: Joel Voelker; Lead Design: Jessica Walsh

6.39 and 6.40: Courtesy of Pentagram Design, Creative Director: Paula Scher/Pentagram

6.41, 6.42, 6.43, 6.44 and 6.45: Agency: LMG Audace & créativité; Creative Direction: Jacques-Dominique Landry, Louis Martin; Art Direction: Mathieu Tremblay; Design: Mathieu Tremblay, Laurent Grislain; Photography: Laurent Grislain; Client: Restos Plaisirs Group

6.47: FedEx service marks used by permission.

6.48 and 6.49: Jennifer Lucey-Brzoza

6.50: Image Conscious Studios LLC

6.51: Trustees of Boston University

6.52 and 6.53: Jennifer Lucey-Brzoza

6.54: Robin Rotman for Desk Plants, DeskPlants.com. 2018. Pages 178-185: Darah Rifkin

6.55: Yuyang Yuan, Yunyi He, Cara Kingsley, Xinyi Ye, Ryan Huff, Heather Hamacek, Morgan Stukes, Anthony Boskinis

7.01: Creative Direction/Design: Stefan Sagmeister; Photography: Timothy Greenfield Sanders

7.02: Agency: LMG Audace & créativité; Creative Direction: Jacques-Dominique Landry, Louis Martin; Art Direction: Mathieu Tremblay; Design: Mathieu Tremblay, Laurent Grislain; Photography: Laurent Grislain; Client: Restos Plaisirs Group

7.03, 7.04, 7.05 and 7.06: Creative Director: Matthew Squadrito, Terry Squadrito

7.07 and 7.08: Courtesy of Pentagram Design, Creative Director: Paula Scher/Pentagram

7.09: Design Ranch

7.10 and 7.11: Image Conscious Studios LLC

7.12, 7.13 and 7.14: Sprout Studios Boston

7.15: Courtesy of Pentagram Design, Creative Director: Abbott Miller/Pentagram

7.16 and 7.17: ©dotdash ©Florain Groehn Photography

7.18, 7.19 and 7.20: Pentagram Design

7.21 and 7.22: Gee Beauty; GJP Advertising – Lisa Greenberg, Trevor Schoenfeld

7.23, 7.24 and 7.25: @SNASK

7.26: Creative Direction: Stefan Sagmeister; Design: Joe Shouldice; Photography: Henry Leutwyler

7.27: Courtesy of Pentagram Design, Creative Director: Paula Scher

7.28: Courtesy of Pentagram Design, Creative Director: Paula Scher
Pages 197-205: Ziqian Liu

7.29 and 7.30: Ziqian Liu

7.31 and 7.32: Nicole Layne

7.33 and 7.34: Lindsay Kerr

7.35 and 7.36: Yuling Lu

RED
bright red: exciting, energizing, sexy
negative: overly aggressive, violent, danger

brick red: earthy, warm, sturdy

deep reds: rich elegant, tasty, expensive, mature

pink: romantic, affectionate, soft, sweet tasting, innocent, youthful
negative: too sweet

bright pink: exciting, playful, hot
negative: gaudy

ORANGE
fun happy
negative: loud, frivolous

terra cotta: earthly, warm, wholesome, welcome, abundance

coral: vital, juicy

YELLOW
joyful, illuminating
negative: hazard

light yellow: cheering, happy, soft

golden yellow: nourishing, wheat, comfort food

BROWN
tan: rugged, outdoor, woodsy

chocolate: rich, appetizing

earth: steady, wholesome, traditional

BLUE
light: calm, clean
negative: aloof, distant, melancholy

sky: calm, cool, true, distance

bright: electric, energy, brisk, flags

periwinkle: genial, lively, cordial

deep blue: credible, authoritative, strong, reliable, loyal, uniforms

GREEN
dark green: nature, trust, refreshing

light: calm, lightweight

bright: fresh, Irish, lively

emerald: luxurious

foliage: natural, hearty, balance, harmony, restful

chartreuse: artsy, bold, trendy
negative: gaudy, tacky, slimy

olive: military, classic, safari
negative: drab

lime: fresh, citrus, tart

aqua: water, cleansing, refreshing

teal: serene, cool, tasteful, confident

turquoise: infinity, compassionate, water, sky, gemstone, tropical, oceans

PURPLE
lavender: romantic, nostalgic, lightweight

mauve: sentimental, wistful, thoughtful

amethyst: protective, peace of mind

blue purples: spiritual, mysterious, enchanting, contemplative

red purple: sensual, dramatic, witty

deep purple: visionary, rich, royal, distant

BLACK
power, authority
negative: in excess it can be intimidating and unfriendly

GRAY
neutral, conservative, security, reliability, creates a background for other colors

WHITE
efficiency, simplicity, fairness, order

Bibliography and Recommended Readings

Publications

Tony Brook and Adrian Shaughnessy, *Paula Scher Works*, Unit Editions, 2017.

Andy Cruz, Rich Roat, Ken Barber, House Industries: *The Process Is the Inspiration*, Watson Guptill Publications, 2017.

Joann Eckstut and Arielle Eckstut, *The Secret Language of Color*, Black Dog and Leventhal Publishers, 2013.

Leatrice Eiseman, *The Complete Color Harmony Pantone Edition*, Rockport Publishing, 2017.

Stephen J. Eskilson, *Graphic Design, A New History*, Laurence King Publishing, 2019.

Michael Evamy, *Logo*, Laurence King Publishing, 2015.

Roger Fawcett-Tang, *Numbers in Graphic Design*, Laurence King Publishing, 2012.

Maitland Graves, T*he Art of Color & Design, Echo Point Books & Media*, reprint edition, 2019.

Phillip B. Meggs and Alston W. Purvis, *Meggs' History of Graphic Design*, Wiley, 2016.

Josef Müller-Brockmann, *Pioneer of Swiss Design*, Lars Muller, 2015.

Stefan Sagmeister and Jessica Walsh, *Sagmeister and Walsh: Beauty*, Phaidon, 2018.

Stefan Sagmeister, et al, *Things I Have Learned in My Life so Far*, Harry N. Abrams Publisher, 2013.

Timothy Samara, *Letterforms: Typeface Design from Past to Future*, Rockport, 2018.

Timothy Samara, *Making and Breaking the Grid*, Rockport, 2017.

Adrian Shaughnessy, *How to Be a Graphic Designer Without Losing Your Soul*, Princeton Architectural Press, new edition, 2012.

Erik Spiekermann, *Stop Stealing Sheep and Find Out How Type Works*, Adobe Press, 2013.

Kassia St. Clair, *The Secret Lives of Color*, Penguin Books, 2016.

Robin Williams, *The Non-Designer's Design Book*, second edition, Peachpit Press, 2014.

Jon Wozencroft, *The Graphic Language of Neville Brody*, Universe, 2002.

Documentaries, Podcasts, and Websites

Abstract, a series of documentaries on Netflix about designers.

AIGA Eye on Design, online magazine: https://eyeondesign.aiga.org/

Chermayeff & Geismar & Haviv, website: https://www.cghnyc.com/

Communication Arts, website: www.commarts.com

Design Matters, Debbie Melman's world's first podcast about design: https://www.designmattersmedia.com/designmatters

Letterpress, type setting video, Maryland Institute College of Art, Globe Collection and Press: http://globeatmica.com/

Pentagram, website: https://www.pentagram.com/

Radio Lab Colors, podcast, 'Why Isn't the Sky Blue?' produced by Tim Howard, WNYC Studios: https://www.wnycstudios.org/podcasts/radiolab/segments/211213-sky-isnt-blue

Sagmeister & Walsh, websites: https://sagmeisterwalsh.com/ sagmeister.com, and andwalsh.com

Stefan Sagmeister, TED Talk, "The power of time off": https://www.ted.com/talks/stefan_sagmeister_the_power_of_time_off?language=en

3 Point Perspective, The Illustration Podcast, Society of Visual Storytelling: https://www.svslearn.com/3pointperspective

Index

Acknowledgments

My gratitude to the wonderful Bloomsbury team: Leafy Cummins, for expertly guiding me through this intense (but fun) year of writing, designs, and edits; Louise Baird-Smith, for championing this edition from the start; and Faith Marsland for her wise counsel and support. Lou Dugdale for designing the gorgeous cover; Roger Fawcett-Tang for the elegant page designs, James Tupper and Ken Bruce for their production expertise; and Rebecca Bigelow for the thorough copyedits.

I am delighted to include designs from Boston to Brisbane—including world-class Pentagram, Chermayeff & Geismar & Haviv, Sagmeister & Walsh, as well as independent studios, designers, and photographers, including Boston-area friends Jennifer Lucey-Brzoza, Joshua Sweeney, and Jason Zucco. Thank you all for your inspiring creativity and generosity.

Many people shared their talents and support during this process. My heartfelt thanks go to my mom Lois, for a lifetime of art lessons, and my late father Francis for encouraging me to write. Big hugs to my family: Michael, Lois, Nancy, and Elaine, Joni, Jeff, Frank, Tom, Alexandra, Kent, Anthony, Jacob, Allison, Sarah, and Micaela. Andy, thank you for all of the edits, advice, laughs, and patience. Mary-Ellen, Mina, Dottie, Eddie, Michael, Tian, Andrea, Eliza, Anita, Claire, George, Michelle, Rita, Thecla, Ellen, Joel, Robin, Jessica, Barr, Anne, and Eric, thank you for your friendship and support. Rich, the readers and I can't thank you enough for your edits. Thank you for all of the support from my Boston University family, especially, Steve Burgay, Amy Hook, Colin Riley and Lee Caulfield who generously provided their expertise. Stephen Hart, Jackie Cimino, Allison Dawes, Miyo Le, and Junning (Sophia) Qin, thank you for fast responses about software, scans, and translations. My dog Archie slept on the floor next to me through months of writing, and I'm grateful he urged me to take many walking breaks.

Warm thanks to my current and former BU students who contributed their work to this project including Raisa Alcoque, Jamie Balkin, Anthony Boskinis, Brian Forte, Heather Hamacek, Lisa Hayward, Yunyi He, Jessica Hudson, Ryan Huff, Liz Kauff, Lindsay Kerr, Cara Kingsley, Wasin Kittiwan, Nicole Layne, Samantha Lee, Hayley Levesque, Min Li, Ziqian Liu, Yuling Lu, Kori (Mausner) Mirsberger, Pieter Melotte, Stephanie Ploof, Darah Rifkin, Robin Rotman, Amanda Saakari, Morgan Stukes, Val Su, Ghewah Taha, Diala Taneeb, Tao Xuzhi, Mitch Weiss, Xinyi Ye, Christina Yin, Raymond Yu and Yuyang Yuan. And special thanks to Graduate Assistant Yuling Lu—xiè xie!

What a pleasure to work with all of you! Wishing you all happiness and success.

Joyce Walsh

Fish and seafood to share

Dedicated to our daughter, Panda.

First published in the United Kingdom in 2018 by
Pavilion
43 Great Ormond Street
London
WC1N 3HZ

This edition published in 2021

ISBN: 978-1-911216-96-4

A CIP catalogue record for this book is available from the British Library.

10 9 8 7 6 5 4 3 2 1

Reproduction by Mission Productions, Hong Kong
Printed and bound by Toppan Leefung, China

www.pavilionbooks.com

Cover design: Elsewhere Studio
Backcover photography: Chris Hewitt

PRAWN
ON THE
LaWN

Fish and seafood to share

Rick & Katie Toogood

PAVILION

Contents

Foreword

The one thing all my favourite seafood restaurants have in common is that the chefs and owners understand seafood to such a level that it is evident in the way every fish is prepared and cooked. This is very much true of Prawn on the Lawn, a gem of a restaurant, run by my friends Rick and Katie. In their first restaurant, they didn't have cooking facilities, so light marinations, cures and raw fish is where these guys started. I loved the idea of a beautiful fish counter you could choose from and then have your selection simply prepared in front of you while you enjoyed a glass of wine – it was brilliant. I think it was due to this restricted kitchen set-up that Rick's food developed the character it has today.

Simple preparation and cooking of seafood takes some skill – the delicate textures of seafood really shine when the balance of flavours are just right. I love the hit of flavour of the POTL som tam crab salad – fresh, hot, fragrant, salty and cooling – a heady mouthful, especially with freshly picked crab that is somehow not overpowered by it all. It amounts to a plate prepared by a chef that has a real connection to what he is cooking – and that's Rick, he just gets it. So, sitting in any of their restaurants, with a whole crab, a plate of crispy fried seafood or some fresh oysters in front of you, is a joy, whether in the city or by the sea.

I think restaurants are a little about food and lot about other things too. When the balance of everything is right, the whole experience is wonderful. Most of the time you can't put your finger on what it is that makes it all work. I believe there is a sort of "umami" in certain restaurants, just as there is in the world of flavor – you love it, but you don't quite know what it is. I feel that sense of wholeness at POTL. The fish is the star of the room – it sits on the counter in all its glory letting you know what the place is all about, whilst the simple decor is personal and comfortable. A short wine list makes things easy, and the service from Katie and her team is good old-fashioned hospitality that makes you feel so welcome.

I really enjoy Rick's company; whenever a few of us foodie people make our annual trip to Italy to enjoy friendship and truffles, it's not long before we are discussing our mutual love of fish and seafood. Rick has always been somewhere new, or has a trip planned to far-off places, to do the hard research all restauranteurs have to do! And I always want to hear about it. I believe it's through travel, eating and life experience that we become better cooks and restaurateurs, and that is certainly true of POTL. The last time I ate there, I could sense Rick had been to Spain; crisp-fried whole red mullet, punchy romesco sauce and clams were on the menu, taking me right back to the sheer pleasure of simple seafood such as I first experienced myself in Barcelona many years ago.

I shall enjoy having POTL on my bookshelf – it's another great book to introduce the home cook to the joys of seafood cookery, to keep the experienced cook on the straight and narrow, and to remind us that it's all about the quality of seafood and simple preparations – deviate from this at your peril!

Mitch Tonks

Prawn on the Lawn

Our restaurant was built around a pretty clear ethos: Katie and I wanted to create a place where we would love to work, a place where our staff would love to serve great food and a place in which our customers would love to spend time. Moreover, we wanted to create an opportunity to share – share our knowledge and our passion for seafood, share where our ingredients come from, share our hospitality with each person that steps through the door, and ultimately, hopefully, share an enjoyable and memorable experience for all.

It feels like a lifetime ago when it all started. After completing a design degree at university, I was lost as to what route to take. A job in China made me realize that a life sat behind a desk wasn't for me. My mum came to the rescue, and asked me to write down the top three things I was most passionate about. Food came out on top, so off I went to work in a restaurant.

My first job was front of house in a restaurant near to where I grew up in Buckinghamshire . When I moved to London, I continued to work front of house, working my way up the ranks to become general manager at a fish restaurant, but I struggled with not being able to do things my way; my design degree was subconsciously at work, redesigning the dining experience. Along the way I met Katie – she was working in the music industry at various record labels in London. She was focused, driven and ambitious, with an equal desire to do things her way. 'My way' and 'her way' soon became our way!

I left my restaurant job and set about finding a site for us to start our own project. We decided that opening a full-blown restaurant was biting off more than we could chew – or afford! We had weeks of viewings all over north London but nothing was suitable. A good mate of mine sent me a picture of an empty butcher's shop he cycled past. We met up with the landlord and couldn't believe our luck; he gave us the start we needed, a rent-free period with staged increases in rent. We'll be eternally grateful to him, for helping us get our dream off the ground. We had a site, now for a name. The only option was to get a couple of bottles of wine, sit in the park, and see what happened. Katie jokingly blurted out 'Prawn on the Lawn', but it took a couple of weeks for us to realize that this was the perfect name.

When we started to work on the shop, we decided it was going to be a fishmonger's where you could sit and have a glass of wine with some oysters or maybe a cooked crab with some crusty bread. We wanted the styling to be reminiscent of a classic fishmonger's, but without the stainless steel, spray the- walls-down feel to it. The shop was to have wine barrels and bar seating to encourage an informal vibe where people could share the space with others. We did everything on a budget: antique shops, reclaimed materials – you name it, we sourced it to get the dream started.

We enlisted the help of everyone we could think of. One of my brothers sorted out the bespoke fish counter, working through the night to help me build it. Next, we needed a logo and branding – its creation is a perfect example of the support we've had from family and friends. The typeface for our initial logo was done by my cousin, the illustration of the prawn in the deckchair by a friend, and it was all put together and styled by my best mate (and best man) who has recently done an amazing rebrand for us (as well as designed the cover of this new edition). Our website was kept up to date and run by my other brother who has since built our new site.

At the beginning, I worked full time in our restaurant with one other member of staff. Katie kept her day job, working evenings and Saturdays front of house. Soon, we found that the 'eating in' side of what we were doing became far more popular than the fishmonger side, so we renovated the basement of the Islington site to fit in more covers. I started experimenting with some sharing-style dishes, using the cooking skills I learned from mum when I was growing up. However, I was restricted in what I could offer, as we only had a retail licence, so we weren't able to cook to order. It forced us to be more creative, which, looking back, helped shape what we do. Without being able to cook to order, I experimented with ceviches, pickles, cures and poaching fish. It was a challenge at first, but I soon enjoyed coming up with new dishes every week; the key note of every dish was to allow the fish to shine. Katie was then able to start at POTL full time. Right from the start we wanted to source from Cornwall. Partly for sentimental reasons, as we both used to go on holiday there, but primarily because of the amazing quality of the produce. I contacted Johnny Murt, who catches our crab and lobster in Padstow. He'd not shipped stuff up to London before, but we managed to sort out the logistics. It meant we had to start visiting Padstow regularly, just to meet up with suppliers, nothing to do with the fact that we loved the place! Each time, the return journey became harder and harder. When a restaurant became available in the town, we didn't even think twice. We wanted to be nearer to where we sourced the produce and thought our concept would sit really well among the other great restaurants in the town. It also meant we now had a fully working kitchen to cook hot food. We've since opened our second restaurant in Padstow, Barnaby's, which follows a similar style as Prawn on the Lawn with small and large sharing plates but incorporates meat and veggie dishes in addition to seafood. More recently we embarked on a new project, Fish Buoys - a business that we started with our good friend Johnny Murt and his partner Cam, during the first lockdown in 2020, supplying our own restaurants, as well as other local places, with sustainable caught, hand-picked crab and lobster meat.

This sort of support and collaboration has carried us through our entire Prawn on the Lawn life. Since the start, we've had people dropping everything and throwing on a POTL shirt to wait tables, wash dishes and run food stalls for us. One of the best examples of this was the summer of 2020 during the Coronavirus pandemic. With government restrictions we were only able to seat 16 people at our Cornish site, so we came up with the bonkers idea of moving the whole operation to a farm just outside Padstow. We spoke to Ross Geach at Padstow Kitchen Garden who said that his parents had a field on their Farm, Trerethen Farm, that was normally used for weddings but due to COVID they had all been cancelled. A couple of phone calls later and with the kind permission from Ron and Sonia Geach, Prawn on the Farm was born. A 50-seater restaurant in the middle of a field with no running water or electricity! That was both a challenge and one of our greatest achievements yet. It would not have been possible without all the help and support of our staff, suppliers and family who worked tirelessly to put up the marquee and get us open in time for the season. It was very successful and we were fully booked all day, every day from the day we opened in July to the day we closed in October. It was a truly amazing experience in a year none of us will ever forget.

So, now we have another opportunity to share in this book some of our favourite and most popular recipes. We never over-complicate our food – fresh ingredients are key. Fish is so versatile, and many of the recipes are interchangeable between different species. The best approach to take with these recipes is to get used to cooking them, and then start experimenting; mix and match the small dishes and the larger sharing plates. The simplicity of the dishes enables you to come up with a cracker of a meal when entertaining. We've added in drinks pairings, too, to give you the complete experience in one book. Cook them up at home, on the beach, or in the garden on your barbecue. Get friends and family around, crack open a nice bottle of wine or a couple of beers, and share.

Sourcing Fish and Seafood

Where our ingredients come from has always been a vital part of what we do. As much as possible, we try to source our seafood as fresh and sustainable as it can be. It's tough these days to be a consumer who wants to purchase fish in a sustainable way – there has been so much conflicting information in the media about what you should and shouldn't buy. The truth is that there are constantly fluctuating quantities of different species all around our coast, so there's no golden rule.

Catching fish at sea is sustainable if the fishery takes into account the effect their activities are having on the ecosystem in which they operate. In a responsible fishery, there are enough fish left to swim away so that they can continue to reproduce in large numbers. So, by all means, catch fish, but always in moderation. A responsible fishing community will always ensure that their fishing causes as little damage to the seabed as possible and will minimize any by-catch. In addition, the fishery should be well organized and managed to ensure that the provenance of each fish can be traced.

To keep things simple, if every person tried as much as possible to buy fish from day boats, and varied what they ate, it would spread the load across all species. Don't be afraid to ask questions at the fishmonger's, from where the seafood was caught, or how it was caught, to what your fishmonger recommends that day. A good fishmonger should know the flavours and textures of the fish he or she is selling, and how best each type should be cooked. There are so many different and amazing fish available, with so many flavours, textures and uses – give them all a try!

Sterilizing Jars

Some of our recipes call for a sterilized jar to store pickled ginger (see p. 43), kalettes (see p. 54) or chutney (see p. 51) in order to extend the shelf life of the food. If your dishwasher has a very hot cycle, you can sterilize your jars and lids in that. Otherwise, we've added a couple of alternative methods below.

First, wash the jars and lids in hot, soapy water, making sure that there is no residue on them, and then rinse thoroughly in hot water. Heat the oven to 160°C fan/180°C/350°F/gas mark 4. Stand the jars, bottles and lids on the oven shelf and leave for 10 minutes to sterilize. Turn the oven off and keep them warm until ready to fill.

Alternatively, wash the jars and lids, then stand them right-side up on a wire rack in a large pan, making sure that they do not touch each other or the sides of the pan. Cover completely with water and then bring to the boil. Simmer for 10 minutes and then remove from the water and stand upside down on a clean, thick cloth to drain. Dry completely in a preheated oven at 90°C fan/110°C/225°F/gas mark ¼, right-side up on a baking sheet for about 15 minutes. They can be kept warm in the oven until required.

Key Marinades and Recipes

In this section, you will find a collection of basic recipes that can be used in a wide variety of ways, and that suit many different types of seafood. We base our menu on the day's catch, which means we have to be very flexible – these key recipes enable us to do that. Master these recipes and you can make anything from soups to marinated fish and roasted shellfish. I would encourage you to play around with different species of fish and shellfish with these recipes. To a certain extent, this ethos can be carried through the whole book – don't be afraid to experiment.

gin
nese Marinade
n Dipping Sau
d Paprika Dip
Crème Fraîch

Vietnamese Marinade

This is one of the most versatile recipes in the book. It can be used across the whole spectrum of fish and shellfish species, both as a marinade and as a dipping sauce. Whilst on our honeymoon in Vietnam, Katie and I went on a motorbike tour of seafood restaurants around Ho Chi Minh City. We enjoyed one particularly amazing dip with some barbecued prawns. When we returned, I tried to recreate it and hope I've done it justice. This makes quite a lot of marinade, but any extra can be kept for up to 3 days in the fridge, or frozen for later use.

Put all the ingredients into a food processor or blender and blitz until as smooth as possible. Don't worry if it looks a little 'bitty', as it will soften down during the cooking process. You now have the basis for an array of different recipes.

Makes about 150ml/5fl oz/scant ⅔ cup (enough dipping sauce for 10)

2 garlic cloves, peeled
a handful of coriander (cilantro), stalks and all
5 kaffir lime leaves
1 thumb-sized piece of fresh ginger, peeled
1 birdseye chilli, roughly chopped
2 lemongrass stalks, topped and tailed, roughly chopped
a splash of Thai fish sauce (nam pla)
juice of 2 limes
50 ml/1¾ fl oz/scant ¼ cup extra-virgin olive oil

Soy Marinade

For Katie's 30th birthday celebrations, a big crowd of us went to stay in an amazing house in Oxfordshire. I'd been playing around with this recipe for a couple of weeks. Since I was in charge of the food and knew that Katie was a massive fan of scallops, I decided to unleash it on the 20-strong group. I cooked the scallops over a fire pit and it went down a storm. I've since tried this marinade in loads of other ways – as a dip as well as a marinade (see the Squid with Tenderstem Broccoli, p. 44, or the Salmon Sashimi, p. 43). It's another versatile recipe to keep up your sleeve.

Makes about 100 ml/3½ fl oz/ 7 tbsp

1 tbsp sesame oil
3 tbsp soy sauce
1 tsp finely chopped garlic
1 tsp finely chopped fresh ginger
juice of ½ lemon

Combine all the ingredients and job done!

Extra-virgin Olive Oil and Lemon Dressing

These two ingredients can transform fish or shellfish from just average into something to remember. It's a combination I remember from sailing around Greece with my parents.

We're known for keeping things simple, but even I wasn't sure about adding this to the book! Please try it though.

The ratio:
70% extra-virgin olive oil (the best you can buy)
30% freshly squeezed lemon juice

Whisk together and that's it!

Nam Jim Dipping Sauce

One of our first ever holidays together was in Thailand. Katie, being obsessed with the country, planned the whole thing. Of all the amazing things we ate, this was one of the key recipes I took from the experience. The name *nam jim* is rather generic in Thailand and covers a variety of sauces, but this one I've tweaked and it works in a variety of ways: as a marinade, as a sauce and as a dip.

Mix all the ingredients together in a bowl and there you go! Purists would say that palm sugar ought to be added, but I prefer to let the natural sweetness of the seafood add this element.

Makes about 6 tbsp (enough dipping sauce for 4)

1 tbsp Thai fish sauce (nam pla)
1 red chilli, deseeded and finely sliced
2 tsp finely chopped fresh ginger
juice of 2 limes

Smoked Paprika Dip

Another great dip for cold cooked shellfish, this also works perfectly with fish tempura.

Just mix it up.

Serves 4

4 tbsp crème fraîche
½ tsp smoked paprika (pimentón) (the hotter the better)
1 tsp lemon juice
a pinch of sea salt and freshly ground black pepper

Garlic Crème Fraîche

I use crème fraîche as an alternative to mayo. At home, when I was growing up, we used to put it with everything and it always seemed to taste good. This is a garlicky version – we use it in the restaurant with a wide variety of seafood, from smoked salmon to crab, or as a dip for all the different crustaceans on a fruits de mer platter.

Serves 4

4 tbsp crème fraîche
½ tsp garlic paste (shop-bought)
2 spring onions (scallions), finely sliced
1 tsp lemon juice
a pinch of sea salt and freshly ground black pepper

Mix all the ingredients together and get dipping!

Shallot Vinegar

A classic accompaniment to oysters; red wine vinegar is traditionally used, but you could also experiment with different flavoured vinegars.

Serves 4

100 ml/3½ fl oz/7 tbsp good-quality red wine vinegar
1 banana shallot, finely chopped

Combine the ingredients and drizzle over fresh oysters.

Cold Tapas

These dishes are our roots. When we first opened, we didn't have a hot food licence, so all our dishes were served cold. The key to this section is preparation. If you're putting on a spread for friends and family, these recipes are all perfect to throw together at the last minute, as long as you've prepared all the components beforehand. The Mackerel Pâté (see p. 51), in particular, is a real crowd pleaser.

d Sa

ts, Capers and

c Crème Fraîch

ing Rollmops

te Wing with

Prawn on the Lawn

This is the first ever dish we served in the restaurant. We had the name and the shop, so then we needed a dish that wasn't too gimmicky. Right from the off, this reflected all the other dishes that followed – bags of flavour, simple to make, it packs a punch and is a true classic.

Scoop the flesh from the avocados into a mixing bowl. Add 2 tbsp of lime juice and mash. Season with salt and pepper, to taste. This mix can be left in the fridge until needed; just leave an avocado stone in the mix and cover in cling film (plastic wrap) and it won't turn brown.

Finely chop the coriander, reserving a few leaves for garnish.

Toast the soda bread, then spoon the avocado mix equally over the slices. Place 3 cooked prawns on each slice. Sprinkle the chilli over the prawns, pour 1 tbsp of the remaining lime juice on top of each slice and sprinkle over the chopped coriander.

Garnish with the reserved coriander, or use the micro-coriander leaves if you have them. Serve immediately, with a wedge of lime.

Serves 4

2 ready-to-eat avocados, Hass are best
6 tbsp lime juice
sea salt and freshly ground black pepper
3–4 sprigs of coriander (cilantro)
4 slices of soda bread (see p. 156)
12 cooked king prawns (jumbo shrimp), peeled and deveined (you can buy raw if you prefer and cook them in boiling lightly salted water for 1–2 minutes)
1 fresh red chilli, deseeded and finely diced
micro-coriander (cilantro), to garnish (if available)
1 lime, quartered, to serve

Pairs well with
Verdejo
Spanish fizz

Smoked Salmon with Shallots, Capers and Garlic Crème Fraîche

Smoking is a traditional way of preserving fish, and in recent years there have been numerous independent smokehouses opening up, so the technique has seen a real resurgence. Try and hunt down a smokery near you, as each one will have a distinct flavour depending on their own curing recipe and the wood they choose to smoke the fish with.

This is a really simple recipe, but then you don't want anything too OTT, as you'd distract from the flavour of the smoked salmon.

Lay the salmon on the serving dish, with a pile of the shallots and capers heaped next to it, and the lemon wedges on the side. Add a ramekin of garlic crème fraîche and serve with the slices of toasted soda or rye bread.

Serves 4

350 g/12 oz smoked salmon, ideally cut vertically (the traditional Scandinavian way)
2 banana shallots, finely diced
3 tbsp capers
lemon wedges
1 quantity Garlic Crème Fraîche (see p. 27)
slices of soda bread (see p. 156) or rye bread, toasted

Pairs well with
Sancerre

Herring Rollmops

Rollmops are such a 'Marmite' dish: some people hate them, but I love them! My auntie is Swedish – on trips over to see my cousins I'd always gorge on all sorts of pickled fish, packed full of dill. Back then, rollmops seemed like an impossible thing to make by yourself; however, they're remarkably simple, as long as you allow at least three days' pickling time.

Lay the herring fillets out on a chopping board, skin-side down, and sprinkle over the chopped dill.

Carefully roll the fillets up, as tightly as possible, and fasten in place by piercing each roll with a cocktail stick. Transfer to a container, making sure the rollmops don't have much space around them otherwise they won't stay submerged.

Add all the pickling ingredients to a saucepan and bring to the boil, then remove from the heat and leave until cool.

Once the pickling mixture is cool, pour it over the rolled herrings, making sure they are all covered. Cover in cling film (plastic wrap) and store in the fridge for at least 3 days.

To serve, place the rollmops on a serving dish, drizzle with a little pickling liquor, and put a ramekin of garlic crème fraîche and some rye bread on the side.

Serves 4

8 herring fillets, pin bones removed
a bunch of dill, finely chopped
8 cocktail sticks

For the pickling mixture:
1 shallot, finely chopped
2 garlic cloves, crushed
a pinch of mustard seeds
3 bay leaves
a pinch of black peppercorns
grated zest of 1 orange
1 tbsp brown sugar
250 ml/9 fl oz/1 cup white wine vinegar
125 ml/4 fl oz/½ cup dry white wine

1 quantity Garlic Crème Fraîche (see p. 27), to serve
4–8 slices of rye bread, to serve

Pairs well with
Lager
Aquavit

Skate Wing with Baby Gem and Smoked Anchovy Dressing

Skate, or ray as it's also known, is more often than not served the classic way with black butter and capers. Here, it's poached and tossed through a simple salad. The smoked anchovy dressing is the key to getting bags of flavour.

Bring a saucepan of lightly salted water to a simmer and add the bay leaves, the 2 crushed garlic cloves and the peppercorns. Gently lower the fish into the water and poach for 7 minutes.

Using a fish slice or slotted spoon, carefully remove the fish from the pan onto a plate – don't worry if it breaks up. While it is still warm, flake the flesh into small pieces – it should naturally split along the fibres. Leave to cool.

Into a food processor or blender, add the anchovies, the remaining garlic clove, the white wine vinegar and lemon juice, and blend, while slowly adding in the olive oil.

Both stages of this recipe can be done the day before, but be sure to keep the skate in the fridge and remove 10 minutes before using, to allow it to come to room temperature.

Mix the shredded lettuce with the skate, shallots, capers and dressing and mix thoroughly. Transfer to a serving plate and sprinkle with the chives and some sea salt.

Serves 4

sea salt
2 bay leaves
3 garlic cloves, peeled – 2 crushed and 1 left whole
a pinch of whole black peppercorns
300 g/10½ oz skate (ray) wing, bone and cartilage removed (your fishmonger can do this for you)

For the dressing:
½ a 30 g/1 oz tin of smoked anchovies (15 g/½ oz drained weight)
a generous splash of white wine vinegar
juice of ½ a lemon
100 ml/3½ fl oz/7 tbsp extra-virgin olive oil

2 baby gem (Boston) lettuces, finely shredded (the finer the better)
2 shallots, finely sliced
a small handful of capers
a small bunch of chives, finely chopped

Pairs well with
Muscadet

Oysters 3 Ways

This is not one for oyster purists, but it perfectly reflects POTL's obsession with flavour. When buying oysters from your fishmonger, ensure that the oysters are firmly closed and do not feel hollow. They should have some weight to them – this means they still have liquid inside, ensuring they are in great shape. Rock (Pacific) oysters are better than natives for this.

Note: The Pickled Cucumber and Dill Dressing should ideally be made 1–2 days in advance.

Mix the diced cucumber with the white wine vinegar. Add the chopped dill with a pinch of sugar and mix well. Store in the fridge until needed.

Shuck (open and detach from their shells) the oysters and place on crushed ice (or a pile of sea salt) on a serving platter – we use a round metal tray in the restaurant.
For the first 4 oysters, sprinkle the desiccated coconut over equally, then add the chilli, followed by the lime juice, and garnish each with a coriander leaf.

For the next 4 oysters, dollop about ½ tsp crème fraîche on each, then add the same amount of keta caviar, finishing off with a good grinding of black pepper.

For the final 4 oysters, spoon over the pickled cucumber dressing and garnish with a dill leaf from the remaining sprig. Serve immediately.

Serves 4

12 oysters
crushed ice or a quantity of sea salt, for presentation

Pickled Cucumber and Dill Dressing:
¼ cucumber, peeled, deseeded if large and watery, and finely diced
25 ml/1 fl oz/scant 2 tbsp white wine vinegar
4 sprigs of dill, leaves only, roughly chopped, plus 1 sprig for garnish
a pinch of caster (superfine) sugar

Coconut, Chilli and Lime Dressing:
4 tsp desiccated (dried grated) coconut
2 fresh red chillies, finely diced
4 tsp lime juice
4 coriander (cilantro) leaves

Crème Fraîche, Keta Caviar and Black Pepper Dressing:
2 tsp crème fraîche
2 tsp keta caviar (salmon roe)
cracked black pepper

Pairs well with
English fizz

Whilst on our travels around Thailand we ate many *som tam*, freshly made in front of us on the streets of the towns and cities, as well as on the beach. It became Katie's favourite dish and something I had to recreate at home. This is a simplified, less spicy version, but it still packs a punch, combining the typical salty, sweet, sour and spicy flavours of the region. Here, I've used freshly picked crab meat, but you could also use diced prawns (shrimp) or even lobster if you want to make it extra fancy!

Using a mandolin, slice the carrots and green papaya into thin matchsticks. Place in a mixing bowl, add the garlic and pour over the nam jim dipping sauce. Mix thoroughly, cover the bowl with cling film (plastic wrap) and store in the fridge for at least 1 hour to allow the flavours to infuse.

When ready to serve, remove the bowl from the fridge and add the mint, coriander, squashed tomatoes, spring onions, and chilli (if using). Mix thoroughly. Transfer to a serving plate, forming a small pile, and top with the crab meat. Sprinkle with the toasted peanuts and serve.

Serves 4

2 large carrots, peeled
1 small to medium green (unripe) papaya, peeled (or green beans, finely sliced)
2 tsp finely chopped garlic
6 tbsp Nam Jim Dipping Sauce (see p. 26)
a small handful of mint leaves
a small handful of coriander (cilantro) leaves
8 cherry tomatoes, halved (or quartered, if large) and squashed slightly
2 spring onions (scallions), finely sliced
1 fresh red chilli, deseeded and finely sliced (optional)
150 g/5½ oz white crab meat (unpasteurized)
a handful of toasted peanuts, crushed

Crab Som Tam Salad

Pairs well with
Verdejo

Salmon Sashimi with Soy, Mooli and Pickled Ginger

Salmon and soy are a match made in heaven. Take some time to source your fish as sustainably as possible and ask your fishmonger for sashimi-grade salmon. Have a go at slicing it yourself, but if you don't feel confident, just ask your fishmonger to do it for you — that's what we're here for! You will need to start at least the day before you want to serve, to make the pickled ginger.

For the pickled ginger, add the slices of ginger to a bowl, sprinkle with the salt and mix thoroughly. Set to one side for 30 minutes. Squeeze any excess liquid from the ginger and transfer to a sterilized jar (see p. 20).

In a small saucepan set over a low-medium heat, gently heat the rice wine vinegar, water and sugar, until the sugar has dissolved. Turn the heat up and bring to the boil. As soon as it has come to the boil, carefully pour the mixture over the ginger slices and seal the jar. Place in the fridge for at least 24 hours.

If your salmon hasn't already been sliced, cut vertically into 1-cm/½-in slices, cover in cling film (plastic wrap) and store in the fridge until needed.

To serve, place the salmon slices on a serving plate, with a tangle of mooli and spring onion on the side. Bring to the table with a ramekin of the soy marinade and a pile of pickled ginger, for guests to serve themselves.

Serves 4

For the pickled ginger:
100 g/3½ oz fresh ginger, peeled and cut into thin slices
1 tsp sea salt
50 ml/1½ fl oz/3½ tbsp rice wine vinegar
50 ml/1½ fl oz/3½ tbsp water
40 g/1½ oz/3 tbsp caster (superfine) sugar

300 g/10½ oz sashimi-grade salmon, skin removed, sliced vertically into 1-cm/½-in slices

½ a mooli (daikon), peeled and cut into matchsticks (enough for a small handful) (alternatively use radishes)
2 spring onions (scallions), finely sliced into strips

1 quantity Soy Marinade (see p. 24)

Pairs well with
A full-bodied white, such as Sancerre

When POTL Islington began the transition from primarily being a fishmonger to becoming a destination to eat super-fresh seafood tapas, this was one of the first dishes on the menu. The components can all be made up in advance and then assembled at the last minute, making it a great dinner party dish. It's also really healthy.

If the tentacles of your squid are thick, cut them lengthways to make them thinner, so they cook evenly.

Bring a saucepan of lightly salted water to the boil. Add the squid and cook for 1½ minutes. Take a piece out to check after 1 minute: if tender, remove the pan from the heat; if not, then cook for a bit longer.

Drain the squid into a colander, and cover with ice to stop the cooking process.

In a separate saucepan, bring more lightly salted water to the boil and add the broccoli. Cook for 1–2 minutes, until tender. Drain and plunge into a bowl of ice-cold water to stop the cooking process. Drain again, and store in the fridge, along with the squid, until needed.

When ready to serve, add the squid and broccoli to a mixing bowl and mix in all the remaining ingredients, except the poppy seeds, making sure they are well combined. Transfer to a serving plate and form a neat pile. Sprinkle with poppy seeds and serve.

Serves 4

300 g/10½ oz squid, cleaned and cut into rings, including the wings and tentacles
sea salt, to salt the cooking water
ice, for cooling
200 g/7 oz tenderstem broccoli (broccolini)
6 tbsp Soy Marinade (see p. 24)
juice of 2 lemons
a small handful of mint leaves
2 fresh red chillies, finely sliced
3 spring onions (scallions), finely sliced
2 pinches of poppy seeds

Squid with Tenderstem Broccoli and Soy Marinade

Pairs well with
IPA
Albariño

Burrata with Chicory, Anchovies and Almonds

Down the road from our Islington restaurant, we are lucky to have an amazing Italian restaurant called Trullo. It was here that Katie and I first had burrata on the recommendation of the manager, and now good friend, Sam James. I've been obsessed ever since. For me, the key is using the best olive oil you can find — it really makes all the difference.

Heat a good glug of olive oil in a frying pan (skillet) over a medium heat. Season the chicory with salt and add to the frying pan with a splash of the white wine vinegar, then cover with a lid. After a minute, remove the lid and flip the chicory over and fry for a further minute.

Place the chicory on a serving plate. Remove the burrata from its packaging and pat dry with paper towels. Place on top of the chicory and gently break open the cheeses. Pour over a generous amount of extra-virgin olive oil and scatter over the anchovies and flaked almonds.

Serves 4

olive oil, for frying
3 red chicory (radicchio), split lengthways
sea salt, to taste
a splash of sweet white wine vinegar
2 burrata or burratina (use buffalo mozzarella if not available)
a generous glug of extra-virgin olive oil
5 salted anchovies
a small handful of flaked almonds

Pairs well with
Grolleau Gris

Smoked Mackerel Pâté

I usually find mackerel pâté pretty boring. So when we put it on the menu in the early days of the restaurant, I wanted to tart it up a bit. Beetroot adds colour and sweetness and the horseradish adds a bit of bite. If you're in a rush, you can always use a bought chutney, but if you have time, it's nice to make your own.

Make the chutney at least a week in advance. It will keep for a few weeks and works really well with cheese or with cold meats, as well as in sandwiches.

Combine all the chutney ingredients in a saucepan. Top up with water, until it just reaches the level of the ingredients. Bring up to a simmer and cook for about 3 hours over a very low heat, stirring at least every 30 minutes, until the ingredients soften, darken and the liquid has evaporated. You will need to stir more frequently the drier the mixture gets. Set aside to cool. You can transfer it to a sterilized jar (see p. 20), if wished. It will keep for 2–3 weeks in the fridge.

For the pâté, flake the mackerel fillets into a mixing bowl and add the rest of the ingredients, stirring vigorously with a fork. Don't mix it in a blender, as you would lose the texture. Once fully combined, transfer to a serving bowl and serve with the chutney and some bread.

The pâté will keep for up to 1 week in the fridge in a lidded container.

Serves 4

For the chutney:
2 large red onions, finely sliced
1 thumb-sized piece of fresh ginger, peeled and finely chopped
2 garlic cloves, finely chopped
3 cooked beetroot (beets), roughly chopped
½ tsp Chinese five spice
4 tbsp soft light brown sugar
100 ml/3½ fl oz/7 tbsp red wine vinegar
100 ml/3½ fl oz/7 tbsp red wine

For the pâté:
4 fillets of smoked mackerel, skin and pin bones removed
2 cooked beetroot (beets) (not the ones in vinegar), cut into small cubes
small handful of parsley, finely chopped
1 tbsp freshly grated horseradish
1 garlic clove, finely chopped
juice of ½ a lemon
2 tbsp crème fraîche

sourdough or soda bread (see p. 156), to serve

Pairs well with
A Provençal rosé
A chilled light red, such as Gamay

Raw Carabineros Prawns with Basil Oil and Orange Zest

If you haven't tried these prawns yet, you haven't lived! They taste incredible – super sweet – and sucking the heads gives a huge amount of flavour. Featured most prominently on menus in Spain and Portugal, they're typically grilled very simply. This recipe adds a little something extra, without detracting from the amazing flavour.

To make the basil oil, bring a saucepan of lightly salted water to the boil. Plunge the basil leaves into the water for 30 seconds, then remove with a slotted spoon straight into a bowl of iced water. Squeeze out any excess water from the basil and transfer to a small blender or spice grinder. Add the olive oil and blend for 2–3 minutes. Pour into a small saucepan, cover and refrigerate for a few hours. After this time, remove from the fridge and gently heat the oil to bring it back to liquid form. Strain through a fine sieve and set aside until ready to use.

Carefully peel the shells from the body of the prawns, leaving the heads attached. Arrange on a serving plate and drizzle with the basil oil. Sprinkle the orange zest over the top and season the prawns with a little sea salt before serving.

Serves 4

For the basil oil:
a small handful of basil leaves
100 ml/3½ fl oz/7 tbsp light
  olive oil

For the prawns:
12 raw sashimi-grade carabineros
  prawns (jumbo shrimp)
zest of 1 orange
sea salt

Pairs well with
A dry sherry, such
as Manzanilla
Fino

Whipped Cod's Roe with Pickled Kalettes

I love smoked cod's roe — it's so under-used as an ingredient, but is well worth having on the table for everyone to have a dip. My good friend Mitch Tonks gave me a great bit of advice: don't worry too much about removing all the skin, that's where a lot of the smoky flavour is. Kalettes, also known as flower sprouts, are a hybrid of a sprout and kale and they have become very fashionable recently, but we've been buying them from our local producer, Padstow Kitchen Garden, for years. They work so well when pickled and cut through the smokiness of the roe.

For the pickled kalettes, place the kalettes in a sterilized sealable jar (see p. 20). Add the remaining ingredients to a saucepan and bring to the boil. Simmer for 1 minute, then carefully pour the mixture over the kalettes and seal the jar. Leave in the fridge for at least 10 hours before serving. They can be kept for up to 1 month in the fridge.

For the whipped cod's roe, remove any thick veins from the skin-like sac with a sharp knife, and place the roe into a food processor or blender. Add the lemon juice, zest and garlic, and blend, while slowly adding the olive oil. Transfer the mixture to a mixing bowl and add the crème fraîche. Using a balloon whisk, whip the mixture to a light texture.

Spoon the mixture into a serving bowl and create a well in the middle. Halve 8–10 of the pickled kalettes and place in the well, then sprinkle with the cucumber, sun-dried tomato, salt and pepper. Serve with flatbread or soda bread.

Serves 4

For the pickled kalettes:
2 large handfuls of kalettes
75 ml/2½ fl oz/5 tbsp white wine
75 ml/2½ fl oz/5 tbsp white wine vinegar
75 ml/2½ fl oz/5 tbsp water
75 g/2½ oz/6 tbsp caster (superfine) sugar
1 sprig of thyme
2 garlic cloves, crushed

For the whipped cod's roe:
200 g/7 oz smoked cod's roe
juice and zest of 1 lemon
2 garlic cloves
150 ml/5 fl oz/scant ⅔ cup light olive oil
200 ml/7 fl oz/scant 1 cup crème fraîche

For the garnish:
thumb-sized length of cucumber, finely cubed
5 sun-dried tomatoes, roughly chopped
2 pinches of sea salt and freshly ground black pepper

flatbread or soda bread (see p. 156), to serve

Pairs well with
A light dry rosé from the Loire Valley or Provence

Sea Trout Ceviche with Avocado

This is a super-simple version of a ceviche. Katie and I went on a fishing trip off the Yucatán Peninsula, Mexico, close to an island called Holbox. As with all fishing trips, I caught nothing, although Katie fortunately caught a few sea trout. We pulled up to a deserted island where I filleted the fish and our fisherman made this very dish for us, there and then. If you can't find sea trout, salmon makes a good alternative.

In a mixing bowl, place the sea trout, red onion, chopped coriander, half the chilli, half the lime juice and a few pinches of salt and pepper. Mix thoroughly and allow to marinate for around 5 minutes, so that the acidity of the lime 'cooks' the fish.

Turn the mixture into a serving dish. Place the avocado on top and garnish with the reserved coriander.

In a separate bowl, combine the remaining lime juice and chilli. Add a generous pinch of salt and pepper and stir. Serve the ceviche with the extra lime juice on the side for those who would like an extra chilli kick to the dish.

Serves 4

350 g/12 oz sashimi-grade sea trout, skinned and diced into 2-cm/¾-in cubes
1 medium red onion, finely diced
1 medium bunch of coriander (cilantro), finely chopped, reserving a few leaves for garnish
2 fresh red chillies (habanero ideally), finely sliced
juice of 8 limes
sea salt and freshly ground black pepper
1 avocado, diced

Pairs well with
Sparkling white wine

Martin Morales, owner of the amazing Ceviche restaurants, really opened my eyes to the process of 'cooking' fish by using citrus. It was just before we opened the Islington branch of POTL that Katie and I ate at his awesome restaurant on Frith Street, London. For us, not having any cooking facilities in the original POTL, this was the perfect way to enhance the flavours of our fresh fish and shellfish without using any heat.

'Tiger's milk' is the Peruvian term for the citrus-based marinade that cures the seafood in a ceviche. In Peru, this invigorating potion is often served in a small glass alongside the ceviche and is believed to be a hangover cure as well as an aphrodisiac.

Using a food processor or blender, blitz all the ingredients for the tiger's milk thoroughly. Pass through a sieve, to remove the pulp, and set the liquid to one side.

Lay the scallop slices out on a serving plate and pour the tiger's milk evenly over the top, making sure each slice is covered. Drizzle the passion fruit seeds over (try to get roughly 1–2 passion fruit seeds on each scallop), sprinkle with the red chilli and garnish with the coriander leaves. Serve immediately.

Serves 4

For the tiger's milk:
1 stick of celery, roughly chopped
1 garlic clove
1 fresh green chilli
juice of 3 small limes
½ a thumb-sized piece of fresh ginger, peeled

9 sustainably sourced scallops, roes removed, thinly sliced into discs
seeds of 1 passion fruit
1 fresh red chilli, deseeded and finely diced
a handful of coriander (cilantro) leaves (use micro-coriander, if you can find it)

Scallop Ceviche

Pairs well with
Champagne
Sparkling Albariño

Octopus Carpaccio

Octopus is one of those species that people are often too frightened to cook. A definite crowd pleaser, this recipe ensures a super-tender meat and, once mastered, you can add your own twists by mixing up different herbs to garnish. Start the day before you want to serve.

In a large stockpot or saucepan, big enough to hold the octopuses, heat a splash of olive oil over a low heat. Add the garlic and cook for about 3 minutes, until softened. Add the octopus and the bay leaves, then fill with enough boiling water to cover the octopus. Bring to a simmer, then cover and cook for 1½ hours. After this time, you can check if the octopus is cooked by squeezing the thickest part of one of the tentacles with tongs. If it feels soft, with little resistance, it's done. The octopus will be very tender and delicate, so carefully remove from the pot, transfer to a chopping board, and set aside.

Lay out 2 large sheets of cling film (plastic wrap), one on top of the other, on your work surface.

While the cooked octopus is still warm, remove each tentacle from the body and lay them lengthways alongside each other in the middle of the sheets of cling film. Lift the cling film over the tentacles and tightly roll into a sausage shape. Pierce a few holes in the cling film, to allow air to escape, then simultaneously twist the cling film at each end to tighten into a firm roll. Wrap in one more layer of cling film, ensuring that the ends are completely sealed. Place in the fridge overnight, or for at least 12 hours, to cool and set. The natural gelatine released from the octopus will set the tentacles together.

Remove from the fridge just before serving. Remove the cling film layers – the sausage shape will hold, but act fast. Lay the sausage on its side on a chopping board and, using a sharp knife, carefully cut thin slices to give you perfect rounds of tender octopus. Lay the slices out on your serving plate/s, drizzle with a little white balsamic vinegar and olive oil, and finally sprinkle with a little salt, parsley, chilli and chives, to garnish. Serve immediately.

Serves 4

a splash of olive oil
3 garlic cloves, crushed
2 whole Mediterranean octopuses, around 2 kg/4 lb 8 oz each, cleaned (frozen Mediterranean octopuses are usually already cleaned. Avoid English octopuses – they are too difficult to cook)
4 bay leaves

To garnish:
a drizzle of white balsamic vinegar or sherry vinegar
a drizzle of good-quality extra-virgin olive oil
sea salt flakes, to taste
2 sprigs of parsley, leaves only, finely chopped
2 fresh red chillies, deseeded and finely diced
4 chives, finely sliced

Pairs well with
Albariño

Seared Tuna with Chilli, Soy and Mirin

Originally, Prawn on the Lawn was only meant to be an oyster bar attached to a fishmonger. We had our Scallop Ceviche (p. 58) and Prawn on the Lawn (p. 30) dishes on from the start, but when customers flocked for the dining side of the concept, I had to start thinking of new recipes. We were sourcing incredible line-caught yellowfin tuna at the time, and it seemed the perfect fish to use – it didn't need to be cooked, as it was sashimi-grade. We brought in a tiny, single induction hob, so that we could sear the fish a little, and it's been on the menu ever since!

You can ask your fishmonger to cut the tuna to size, so that you have chunks that are roughly 10-cm/4-in thick, by 20-cm/8-in long.

Heat a large, non-stick frying pan (skillet) over a medium heat and add a drizzle of olive oil. Place the tuna in the pan carefully. As the meat cooks it will change from a deep red to a grey colour and you only want to see it colour just a few millimetres in (which will take about 20 seconds). Roll the tuna over and repeat the process until all the sides are seared. The ends should remain uncooked.

Remove from the pan to a plate and allow to cool. Store in the fridge for at least 30 minutes, for the tuna to firm up.

Using a sharp knife, cut the tuna loin into roughly 1-cm/½-in thick slices. Fan the slices out on a serving plate and sprinkle over the chilli, spring onion and coriander. In a small dish, mix the soy sauce and mirin together and place next to the tuna, to serve. You can dip the tuna slices into it, or pour it all over the tuna, it's up to you!

Serves 4

olive oil, for searing
500 g/1 lb 2 oz sashimi-grade, line-caught tuna loin
2 fresh red chillies, deseeded and finely diced
4 spring onions (scallions), finely sliced
a small bunch of coriander (cilantro), leaves only, finely chopped, or micro-coriander
75 ml/2½ fl oz/5 tbsp soy sauce
2 tbsp mirin (rice wine vinegar)

Pairs well with
Chinon
Pinot Noir

Langoustines are an incredible species; their flavour is super sweet. There's no getting away from the fact that the fresher they are, the better they taste, so avoid previously frozen and opt for live langoustines – they also have a much firmer texture. In this recipe, the extra ingredients are there to enhance the flavour, but it is the langoustines that do the talking. This recipe can be served cold or hot.

If you are buying ready-cooked langoustines, ensure that they haven't been previously frozen, as the meat can be quite mushy. The best option is to buy them live and either boil in salted water for 2 minutes or, if you can handle it, split them in half live before cooking them. To do this, place the point of a sharp knife on the cross on the top of the langoustine's head. Cut firmly down through the head to split. Turn the langoustine, hold firmly and continue cutting through the centre line, down the full length of the body. Open out, pull out the 'sandbox', near the head, and devein.

If the langoustines were previously cooked, they can be served right away. Lay the split langoustines on a serving plate, meat-side up, place a sprig of thyme on each half and drizzle with olive oil. Serve with the lemon halves (charred in a hot pan, if wished). Taste before seasoning with salt, as they may not need it.

If you want to serve them hot, follow these further steps.

Heat a little olive oil in a frying pan (skillet) over a medium-high heat. Sprinkle the langoustines with the thyme sprigs and place in the pan, meat-side down, ensuring they're sitting on top of the thyme sprigs. Add the lemon halves to the pan, cut-side down. Fry for 1 minute, then turn the langoustines over and fry for a further 30 seconds. Transfer to a serving plate, sprinkle a little sea salt over them and garnish with the charred lemon halves.

Serves 4

16 small to medium or 8 large langoustines
32 sprigs of fresh thyme or 16 if only using 8 langoustines,
extra-virgin olive oil, for drizzling
1 lemon, halved
sea salt, to taste

Split Langoustines with Thyme and Olive Oil

Pairs well with
Prosecco
A dry sherry, such as Manzanilla
Fino

Beetroot-cured Salmon with Tarragon Ricotta, Pickled Cucumber and Dill

Curing was a method of preserving fish and meat back in the days when fridges weren't around. It's a great way of changing the texture and adding a huge amount of flavour to the fish, especially when other ingredients are introduced to the curing process. Beetroot adds a sweet, earthy taste as well as a splash of purple to the salmon.

Place the salmon fillet in a non-reactive (glass, ceramic or stainless steel) container.

Thoroughly mix the salt, sugar, beetroot, lemon zest and juice together. Pour over the salmon fillet,

cover with cling film (plastic wrap) and refrigerate for 8 hours.

Once the salmon has firmed up, remove from the fridge, wash off the curing salt and pat dry with paper towels. Cut the salmon vertically into 1-cm/½-in slices (first cut down to the skin, then slice horizontally, to remove each slice cleanly from the skin) and lay on a serving plate.

Thoroughly mix the pickled cucumber ingredients together and pile next to the salmon. Do the same with the tarragon ricotta ingredients. Serve with slices of rye or soda bread.

Serves 4

300 g/10½ oz fresh salmon fillet, skin on
50 g/1¾ oz/3 heaped tbsp table salt
50 g/1¾ oz/3 heaped tbsp caster (superfine) sugar
2 cooked beetroot (beet), finely chopped
zest and juice of 1 small lemon

For the pickled cucumber:
½ cucumber, peeled and cut into 1-cm/½-in cubes
a small handful of chopped dill
a good splash of white wine vinegar

For the tarragon ricotta:
100 g/3½ oz ricotta
4 sprigs of tarragon, leaves only, finely chopped
zest of 1 lemon

rye bread or soda bread (see p. 156), to serve

Pairs well with Prosecco

Razor Clams with Nam Jim, Coconut and Passion Fruit

Whilst out in Thailand, I became obsessed with fish sauce — it has an incredible flavour when mixed with other classic Asian ingredients such as the ginger and lime juice in the nam jim dipping sauce. In this recipe it really brings out the fresh meatiness of the clams.

Razor clams are my favourite of this family of shellfish. They should be bought live from the fishmonger — look for clams that are tightly closed (they might hang out of the end of the shell, but should instantly retract when touched). Although there's a little work involved to clean the clams, the dish can be prepared in advance — you can cook and prepare the meat of the razor clams the day before and store, covered, in the fridge. It's a great dish for entertaining — it's packed with flavour and looks amazing on the plate.

Place the razor clams in a bowl, fill it with water to submerge them completely, and leave in the fridge for 15 minutes. This will remove any sand or grit in the meat. Drain.

Heat a large, lidded saucepan over a medium-high heat. When hot, add the razor clams and 50 ml/1½ fl oz/3½ tbsp of water and quickly cover with the lid. Steam for about 2 minutes, until the razor clams pop open and the meat has detached from the side of the shells. Remove the pan from the heat and tip the clams into a colander placed in the sink, covering them with ice to stop the cooking process.

Meanwhile, make the nam jim dipping sauce and set aside.

To prepare the razor clams, remove the cooked meat from the shells and set the shells to one side. Using a sharp knife, remove the gut from the clams (it's the dark brown part in the middle) and discard. Cut the remaining meat into 1-cm/½-in pieces. Mix the clam meat in a bowl with the diced tomatoes, spring onions and passion fruit seeds and 3 tbsp of the nam jim dipping sauce.

Lay out the clam shells on a serving dish and spoon the clam mixture back into the shells. Sprinkle with the coconut, coriander and red amaranth, and serve.

Serves 4

1 kg/2 lb 4 oz razor clams
ice, for cooling
10 cherry tomatoes, finely diced
2 spring onions (scallions), finely sliced
1 tsp passion fruit seeds
½ quantity Nam Jim Dipping Sauce (see p. 26), about 3 tbsp

For the garnish:
2 tbsp grated fresh coconut or desiccated (dried grated) coconut
a small handful of coriander (cilantro), finely chopped, or micro-coriander
a sprinkling of red amaranth micro-herbs, if available (you can source these online or some specialist greengrocers stock them)

Pairs well with
Pinot Gris

Hot Tapas

We've kept the same approach to hot food as we had to the cold food of our earlier years, keeping things simple, but making sure that everything has that punch of flavour. Don't be shy about trying out the recipes with different species of fish or shellfish; there aren't any official rules on what goes with what, so have a bit of fun.

...sh C...

Crab Cakes wit...

...ole a la Planch...

...with Feta, Wit...

...uid with 'Nduje...

Monkfish Cheeks with Mushroom Broth and Seaweed

Whilst we were renovating our Padstow restaurant, I was lucky enough to be given the opportunity to work in Paul Ainsworth's kitchen at No. 6, just around the corner from our place. It was really eye-opening – the attention to detail, the passion and the organization – I was inspired. One awesome technique I picked up was the use of Marmite, mixed with a little butter, to add richness to the meat that it is brushed on. The most suitable fish I found to use this technique on is monkfish. If you can't find monkfish cheeks, the recipe works just as well with a similar quantity of monkfish fillet.

Preheat the oven to 160°C fan/180°C/350°F/gas mark 4.

Place the dried porcini mushrooms in a bowl, pour over the warm water and set aside.

In a large saucepan, warm a splash of olive oil over a medium heat, add the shallots and garlic and cook for about 10 minutes, until softened.

Add the thyme sprigs and the porcini mushrooms, including their soaking water. Bring to the boil, then lower the heat, add the fresh mushrooms and leave to simmer for 15 minutes.

Place the monkfish on a plate, drizzle with olive oil and season with salt and pepper on both sides.

In a small saucepan, gently melt the butter and mix in the Marmite. Once combined, pour into a warmed ramekin and set aside.

Heat a little olive oil in a griddle or frying pan (skillet) set over a medium-high heat. Add the monkfish and fry for 2 minutes on each side. Transfer the fish to a roasting pan and baste with the buttery Marmite on both sides. Roast in the hot oven for 5 minutes, then remove and allow to rest for 3 minutes.

Ladle the mushroom broth equally into 4 wide serving bowls. Divide the roasted monkfish cheeks between the bowls, and crumble over some dried seaweed, to garnish.

Serves 4

30 g/1 oz dried porcini mushrooms
700 ml/24 fl oz/3 cups warm water
olive oil, for cooking and drizzling
4 shallots, finely sliced
2 garlic cloves, finely chopped
4 thyme sprigs
200 g/7 oz mixed fresh mushrooms, roughly chopped
600 g/1 lb 5 oz monkfish cheeks, cut into 4 or 8 pieces
sea salt and freshly ground black pepper
50 g/1¾ oz/3½ tbsp unsalted butter
1 tsp Marmite (yeast extract)
dried dulse seaweed, to garnish

Pairs well with
Pinot Noir

Thai Crab Cakes with Sweet Chilli Dipping Sauce

We're so lucky to have amazing crab available to us in Padstow (and also in London, as we send it up to the restaurant overnight from the coast). There's nothing better than a whole crab, with some great bread and aioli — but admittedly, it's messy and very labour-intensive. These crab cakes are a great way of eating crab without feeling as though you need a bath after your meal!

To make the sweet chilli sauce, blend the garlic, chillies, sugar, water, vinegar and salt to a rough purée in a food processor. Transfer to a saucepan and bring to the boil. Reduce the heat and simmer for about 3 minutes, until the mixture thickens. Combine the cornflour and water to make a paste and add to the saucepan with the chilli mixture. Bring to the boil and cook for about 1 minute. Leave to cool, then store in the fridge until needed.

Preheat the oven to 160°C fan/ 180°C/350°F/gas mark 4.

Combine all the crab cake ingredients (except the olive oil) in a large bowl and mix thoroughly. Shape into 8 evenly sized balls, then flatten each slightly between the palms of your hands to make a patty.

Heat a large, ovenproof, non-stick frying pan (skillet) over a medium heat. Add a drizzle of olive oil and fry the crab cakes for 2 minutes on each side, until golden. Be careful not to break them up. Transfer the frying pan to the hot oven for 5 minutes, to cook the cakes all the way through.

Remove from the oven and transfer to a serving plate. Scatter the spring onions over the top and serve with the lime wedges on the side, for squeezing over the top. Serve with the sweet chilli sauce.

Pairs well with
Riesling from Alsace, Australia or New Zealand

Serves 4 (makes 8 cakes)

For the sweet chilli sauce:
3 garlic cloves, peeled
2 medium fresh red chillies, stalks removed, seeds intact
170 g/6 oz/generous ¾ cup caster (superfine) sugar
180 ml/6 fl oz/¾ cup water
60 ml /2 fl oz/4 tbsp white wine vinegar
½ tsp table salt
1 tbsp cornflour (cornstarch)
2 tbsp water

For the crab cakes:
200 g/7 oz white crab meat (unpasteurized)
75 g/2½ oz/1 cup panko breadcrumbs
1 fresh red chilli, finely sliced
4 spring onions (scallions), finely sliced
4 tbsp chopped coriander (cilantro)
2 garlic cloves, finely sliced
2 tsp finely chopped fresh ginger
zest of 1 lime
75 g/2½ oz fine green beans, finely sliced
1 tbsp crème fraîche
1 tbsp Vietnamese Marinade (see p. 24)
1 egg
olive oil, for cooking

For the garnish:
2 spring onions (scallions), finely sliced
1 lime, cut into wedges

Sand Sole
a la Plancha

After our first year of the Padstow restaurant being open, we thought it would be awesome to take the guys from both Islington and Padstow on a staff trip to Barcelona. It was a great chance to spend some time together and gain a bit of inspiration. We had so many great meals, but one that stood out was when we perched at a bar and had all sorts of seafood simply cooked on the plancha (hot plate) with a drizzle of parsley oil. That, combined with some ice-cold white wine, was one hell of a lunch!

Lay the fish out on a tray, drizzle a little oil on both sides and season well with salt and pepper.

Add the extra-virgin olive oil, parsley and garlic to a food processor, blend for 2–3 minutes and set to one side.

Heat a frying pan (skillet), large enough to hold the 2 fish, over a medium heat and add a little olive oil. Lay the fish in the pan and fry for 4–5 minutes. Flip the fish over and cook for a further 4–5 minutes. If you are using lemon sole, or any other chunky flat fish, you may need to additionally roast the fish in the oven for a few minutes at 160°C fan/180°C/350°F/gas mark 4, to cook through.

Once cooked, transfer to a serving dish and generously drizzle the parsley oil over the fish. Squeeze over fresh lemon, to taste.

Serves 4

2 x 250–300 g/9–10½ oz sand sole, skinned (alternatively use Dover sole, lemon sole or megrim)
olive oil, for drizzling
sea salt and freshly ground black pepper
100 ml/3½ fl oz/7 tbsp good-quality extra-virgin olive oil
a small bunch of parsley, roughly chopped
1 garlic clove, peeled
lemon wedges, to serve

Pairs well with
Albariño

Cod with Feta, Wild Garlic and Pine Nuts

Cheese and fish are ingredients that you wouldn't immediately associate with each other. However, feta, with its salty-sour flavour, works fantastically well when combined with the fresh herbs and wild garlic. We pick our wild garlic from a shaded walled area close to the beach in Padstow, and near to my grandparents' house in London. It's only in season for a couple of months from March onwards.

Preheat the oven to 160°C fan/ 180°C/350°F/gas mark 4.

Heat an ovenproof frying pan (skillet) over a medium-high heat. Add the butter. Score the skin of the cod fillets, drizzle with a little of the olive oil and season. Place the fillets in the pan, skin-side down, and fry for 2–3 minutes to crisp up the skin. Turn the fillets over and fry for another minute.·

Transfer the frying pan to the hot oven and roast the fish for 8 minutes.

Meanwhile, combine the remaining olive oil, feta, wild garlic, basil, mint, dill, lemon juice, pine nuts and white wine vinegar in a bowl. Season well and mix thoroughly.

Remove the frying pan from the oven, transfer the fish to a serving dish, and pour the cheese dressing over the fish to serve.

Serves 4

1 tsp unsalted butter
4 x 200 g/7 oz cod fillets, skin on
100 ml/3½ fl oz/7 tbsp good-quality extra-virgin olive oil
sea salt and freshly ground black pepper
100 g/3½ oz feta cheese, crumbled
a small bunch of wild garlic, leaves roughly chopped (keep some of the flowers if you can, for garnish)
a small bunch of basil, roughly chopped
a small bunch of mint, roughly chopped
a small bunch of dill, roughly chopped
juice of ½ lemon
2 tsp pine nuts, toasted
a splash of white wine vinegar

Pairs well with
Sancerre
Pouilly-Fumé

This is one of the most popular dishes at POTL. 'Nduja is a spicy sausagemeat from Calabria, in southern Italy. We buy ours from a small producer who makes it in Dorset. The recipe works amazingly well with squid. We cross-score the flesh, which allows the squid to cook more quickly, as well as allowing the flavours to penetrate the meat further.

Heat a saucepan over a medium heat and add the 'nduja, butter and lemon juice. Heat until the butter has melted, then thoroughly mix the ingredients together in the pan. You can keep this mixture, refrigerated, for up to a week – you will need to mix it intermittently as it cools and sets. It can be reheated in the pan when needed.

Place the squid pieces in a bowl and pour over half of the melted 'nduja mixture. Mix well, ensuring the squid is covered thoroughly in the 'nduja butter.

Heat a large, non-stick frying pan (skillet) over a high heat and drizzle in a little olive oil. Place the squid tentacles in the pan first, as they take a little longer to cook and don't be tempted to move the squid. After a minute or so, add the scored body and wings, placing the scored side down. The scored pieces will curl up; when this happens, turn the tentacles over along with the wings and scored body, and cook for a further 2 minutes.

Place the squid in a serving dish. Pour over any cooking juices left in the pan, as well as the remaining 'nduja butter, and garnish with the tarragon leaves. Serve immediately.

Serves 4

125 g/4½ oz 'nduja sausagemeat
125 g/4½ oz/½ cup unsalted butter
juice of ½ lemon
4 medium squid, cleaned, body scored and cut into pieces, tentacles and wings included
olive oil, for frying
3 sprigs of tarragon, leaves only

Squid with 'Nduja and Tarragon

Pairs well with
Verdejo

Roasted Scallops with Vietnamese Marinade and Peanuts

This dish utilizes the shells of the scallops. It's simple to prepare, but great for showing-off to mates (or the in-laws!). If possible, source hand-dived scallops – although they are more expensive, they have little impact on the ocean and aren't treated with any preservatives or chemicals, as dredged scallops can be. You could replace the scallops with oysters; simply shuck the oysters and follow the same process.

Preheat the oven to 160°C fan/ 180°C/350°F/gas mark 4.

Arrange the scallops in their shells in a roasting pan. Mix the Vietnamese marinade and the coconut milk together and spoon this equally over the scallops. Sprinkle half the sliced spring onion over the scallops and roast in the hot oven for 12 minutes.

Once cooked, transfer the scallops in their shells onto a serving dish. We sit the shells on a bed of seaweed, but if this isn't available then some finely sliced baby gem (Boston) lettuce leaves could be used to hold the shells in place. Sprinkle the toasted peanuts over, followed by the remaining spring onion, and garnish with the coriander leaves.

Serves 4

4 scallops, in the half shell
4 tbsp Vietnamese Marinade (see p. 24)
2 tbsp coconut milk
2 spring onions (scallions), finely sliced
a small handful of toasted peanuts, crushed
8 coriander (cilantro) leaves

Pairs well with
A fruity German Riesling
New Zealand Sauvignon
Torrontés

Tempura Pollock with Curry Sauce and Crispy Capers

Pairs well with
IPA
Australian Riesling

Serves 4

Pollock has always been known as a cheaper alternative to cod – it is not quite as flakey, but I do love the flavour. This recipe is great for any white fish: plaice, haddock or whiting. You can't go wrong when you tempura fish.

To make the curry sauce, set a saucepan over a medium heat, drizzle in a little olive oil, add the shallots, chillies and garlic, and cook until softened. Set aside.

In a separate dry frying pan (skillet), set over a medium heat, toast the coriander, cumin and fennel seeds and cardamom pods for about 1 minute. Transfer to a spice grinder or pestle and mortar and grind to a powder.

Put the saucepan with the softened shallots, chillies and garlic back over a low heat and add the ground spices. Add in the cinnamon and turmeric and mix thoroughly. Add in the chopped tomatoes and coconut milk and simmer for 15 minutes. Remove from the heat and blend the mixture to a sauce. Pass the sauce through a sieve to remove any bits. If the sauce is too runny, you can return it to the heat for a few minutes, to reduce.

In a small frying pan (skillet) set over a medium heat, add a little olive oil and fry the capers for a couple of minutes, until crispy. Remove to some paper towels and pat dry. Set aside.

For the batter, combine the dry ingredients in a mixing bowl and mix thoroughly. Slowly add the chilled sparkling water, whisking thoroughly until you achieve the consistency of double (heavy) cream.

Ideally, use a deep-fat fryer (if you haven't got one, use a heavy-based saucepan) and heat the vegetable oil to 190°C/375°F. Test the temperature with a kitchen thermometer or when a cube of bread dropped in the oil sizzles immediately.

Coat the pollock strips in the batter mix and fry in batches for about 1 minute, until they are just about to turn golden. Remove with a slotted spoon and transfer to some paper towels to absorb any excess oil, and season.

Pour the curry sauce into a serving dish, pile the tempura on top, and garnish with the crispy capers, spring onion, coriander leaves and lime wedges.

For the curry sauce:
olive oil, for frying
2 shallots, finely sliced
2 bird's eye chillies, finely chopped
4 garlic cloves
2 tsp each of coriander seeds, cumin seeds and fennel seeds
1 tsp black cardamom pods
1 tsp ground cinnamon
2 tsp ground turmeric
1 x 400 g/14 oz can of chopped tomatoes
1 x 400 ml/14 fl oz can of coconut milk

For the crispy capers:
olive oil, for frying
3 tbsp capers

For the tempura batter:
75 g/2½ oz/½ cup plain (all-purpose) flour
50 g/1¾ oz/½ cup cornflour (cornstarch)
1½ tsp baking powder
approx. 150 ml/5 fl oz/scant ⅔ cup chilled sparkling water

vegetable oil, for frying
400 g/14 oz pollock, skinned and cut into thin strips
sea salt and freshly ground black pepper
2 spring onions (scallions), finely sliced, to garnish
a few coriander (cilantro) leaves, to garnish
lime wedges, to garnish

Steamed Mussels and Cockles in a Coconut and Lemongrass Broth

Our Vietnamese marinade packs a real flavour punch on its own, but when you combine it with the juices released from the shellfish it creates a tasty broth. Grab some crusty bread to mop up all those juices. You could also use palourde clams, razor clams and/or even throw some prawns (shrimp) in when steaming, to add an extra element to the recipe.

Choose to purchase live mussels and cockles. The mussels should have securely closed shells. If they are open, tap them on a hard surface; if they close they are still alive and fine to eat. The shells of the mussels should be damp and shiny, although this is less noticeable with cockles. Avoid mussels and cockles with broken, cracked or split shells – discard these before cooking.

Mix the Vietnamese marinade and coconut milk together in bowl.

Heat a large, lidded saucepan over a high heat. Once hot, add the mussels and cockles, pour the coconut milk mix over the molluscs and cover with the lid. Cook for about 3–4 minutes, shaking from time to time, until all the molluscs have popped open. Discard any that have remained closed.

Divide between 4 bowls, with an equal amount of broth in each, or transfer to one large serving dish. Garnish with the chopped spring onions, mint and coriander leaves and serve with wedges of lime. Some soda bread or a crusty loaf is also great to have to hand, for dunking in that broth.

Serves 4

6 tbsp Vietnamese Marinade (see p. 24)
200 ml/7 fl oz/scant 1 cup coconut milk
1 kg/2 lb 4 oz live mussels
600 g/1 lb 5 oz live cockles

For the garnish:
4 spring onions (scallions), chopped
a small handful of mint leaves
a handful of coriander (cilantro), roughly torn
4 lime wedges

soda bread (see p. 156) or crusty bread, to serve

Pairs well with
Albariño

Prawn and Oyster Mushroom Noodle Soup

A quick and easy dish; your mates will think you've spent years living in the Vietnamese countryside mastering the art of a mega-tasty, warming soup!

Heat a saucepan over a medium heat, add the coconut oil and red onion and cook for a few minutes to soften. Add the mushrooms and, once these have softened a little, add the Vietnamese marinade and stir in thoroughly. Add the stock, coconut milk and boiling water.

Bring up to the boil, then add the noodles. Reduce to a simmer and cook for 3 minutes.

Finally, add the prawns and the pak choi. If using raw prawns, cook until they have changed from a grey colour to pink; if cooked, just allow a minute for the prawns to heat through.

Transfer to individual bowls, or one large serving bowl, and scatter the garnishes over the soup, to serve.

Pairs well with
New Zealand
Pinot Noir

Serves 4

2 tsp coconut oil or olive oil
2 red onions, finely sliced
100 g/3½ oz oyster mushrooms (or a mixture of different mushrooms), roughly chopped
8 tbsp Vietnamese Marinade (see p. 24)
350 ml/12 fl oz/1½ cups fish or vegetable stock
250 ml/9 fl oz/1 generous cup coconut milk
500 ml/17 fl oz/2 generous cups just-boiled water
150 g/5½ oz fresh wholewheat noodles
16 tiger prawns (shrimp), raw or cooked, head and shell on or off, whatever you prefer
2 small or 1 large pak choi (bok choy), quartered

To garnish:
4 spring onions (scallions), finely sliced
1 fresh red chilli, deseeded and sliced
a small handful of coriander (cilantro), chopped
a small handful of mint, chopped

North African Spiced Mackerel

Martin Murt is one of the local fishermen we work with and when he brings mackerel into our Padstow restaurant, this dish goes straight on the menu. It's hugely popular as mackerel is actually caught just outside Padstow, so you really can't get much fresher.

Preheat the oven to 160°C fan/ 180°C/350°F/gas mark 4. Alternatively, bring a barbecue up to temperature.

Using a sharp knife, score the mackerel along both sides of the body and season with a little salt.

Heat a heavy-based frying pan (skillet) over a medium heat and toast the coriander and cumin seeds until fragrant (about 1 minute), then transfer to a spice grinder or pestle and mortar and crush. Add the crushed spices to a food processor, along with the garlic, paprika, preserved lemon, a splash of the lemon-preserving liquor, parsley, chopped coriander, tomato purée and olive oil, and blend for 1 minute.

Line a roasting pan with some greaseproof (wax) paper and place the mackerel on top. Rub the spice paste over the mackerel and into the scored flesh.

Roast in the hot oven for 8 minutes. Alternatively, cook over the hot barbecue for about 4 minutes on each side.

Transfer the fish to a serving plate, and garnish with the reserved coriander leaves and lime wedges.

Serves 4

4 mackerel, gutted and cleaned
sea salt
1 tbsp coriander seeds
1 tbsp cumin seeds
4 garlic cloves, peeled
2 tbsp paprika
1 preserved lemon, plus a splash of the liquor from the jar
1 small bunch of parsley, roughly chopped
1 small bunch of coriander (cilantro), roughly chopped, with a few leaves reserved for garnish
1 tsp tomato purée (tomato paste)
40 ml/1¼ fl oz/2½ tbsp olive oil
lime wedges, to serve

Pairs well with
A fresh, cold lager
A light wheat beer
A spicy red wine, such as Rioja

Our supplier, Johnny Murt, often has rockling come up in his crab and lobster pots. It's almost unheard of to see it in a restaurant or fishmongers, but I find it suits a tempura batter perfectly – the meat is soft and this marries beautifully with the crunch of the batter. If you can't get hold of rockling, this recipe works just as well for cheaper fillets of fish, such as plaice, coley or whiting.

Mix the plain flour, cornflour and baking powder together, then slowly add the chilled sparkling water, whisking as you go. You're aiming for the consistency of double (heavy) cream and you may need to use less or more water than the quantity listed.

Heat the vegetable oil in a pan or wok deep enough to submerge the strips of fish. You can test when it has reached 190°C/375°F with a kitchen thermometer or when a cube of bread dropped into the oil sizzles immediately. Alternatively use a deep-fat fryer.

Dust the strips of fish in a little flour, then coat them in the batter mix and carefully lower into the hot oil. You may need to do this in 2 batches. Once golden, remove with a slotted spoon and drain on paper towels.

Transfer to a serving plate. Sprinkle with the sliced chilli, mint and spring onions and serve immediately, with the nam jim dipping sauce on the side.

Serves 4

75 g/2½ oz/½ cup plain (all-purpose) flour, plus a little for dusting
50 g/1¾ oz/½ cup cornflour (cornstarch)
2 tsp baking powder
approx. 150 ml/5 fl oz/scant ⅔ cup chilled sparkling water (you may need more or less)
vegetable oil, for frying
300 g/10½ oz rockling or other white fish, skinned and cut into thin strips
1 fresh red chilli, thinly sliced
a small handful of mint leaves
2 spring onions (scallions), finely sliced
100 ml/3½ fl oz/7 tbsp Nam Jim Dipping Sauce (see p. 26), to serve

Rockling Tempura with Nam Jim

Pairs well with
Sharps Cornish Pilsner Lager

Octopus with Coriander Butter Beans and Mojo Verde

Octopus isn't the easiest thing to cook, but following this recipe should ensure a melt-in-the-mouth texture. Combine this with the punch of flavour from the spicy coriander beans and you've got a match made in heaven!

Heat a splash of olive oil in a pot large enough to hold the octopus, over a low-medium heat. Add the garlic and soften for a minute, then add the bay leaves and the octopus and cover with a lid. Cook for 30 minutes. After this time, top up with boiling water, ensuring the octopus is completely covered. Cook for a further 1 hour, then remove the octopus and allow to cool.

In a separate saucepan, heat another splash of olive oil over a medium heat and add the shallots, garlic and chilli. Sweat until softened, then add the butter beans, white wine and stock and keep on a low simmer for 20 minutes to reduce slightly.

In a spice grinder or pestle and mortar, add a small bunch of coriander with a drizzle of olive oil and grind to a paste. Add this mixture to the butter beans and stir thoroughly. Season to taste.

To make the mojo verde, add all the ingredients to a food processor and blend for 2 minutes to a fine paste.

Once cool, cut the head off the octopus and discard, then separate the individual tentacles. Drizzle a little olive oil over the tentacles and season well.

Heat a large, non-stick frying pan (skillet) over a medium heat. Add a little oil to the pan, then add the tentacles, cooking for about 1 minute on each side.

To serve, spoon some of the butter beans onto each plate, top with the tentacles, then spoon the mojo verde over the tentacles. Garnish with the reserved coriander leaves.

Pairs well with
A dry, crisp, lemony white wine, such as Albariño

Serves 4

olive oil
3 garlic cloves, crushed
2 bay leaves
1 x 2 kg/4 lb 8 oz Mediterranean octopus, cleaned (frozen Mediterranean octopuses are usually already cleaned. Avoid English octopuses – they are too difficult to cook)

For the butter (lima) beans:
olive oil
3 banana shallots, finely sliced
6 garlic cloves
1 green bird's eye chilli, finely sliced
800 g/1 lb 12 oz canned butter (lima) beans, drained
100 ml/3½ fl oz/7 tbsp white wine
250 ml/9 fl oz/1 generous cup chicken stock
a small bunch of coriander (cilantro), a few leaves reserved to garnish
sea salt and freshly ground black pepper

For the mojo verde:
a small bunch of coriander (cilantro), blanched for a few seconds in boiling water
2 garlic cloves
juice of ½ a lime
½ tsp cumin seeds
75 ml/2¼ fl oz/4½ tbsp extra-virgin olive oil
generous pinch of sea salt

We are so lucky to have Porthilly oysters growing in the estuary a stone's throw from the restaurant in Padstow. Tim and Luke Marshall care so much about their product. During the busy times of the season we sometimes run out — one call to the guys and they shoot over in their boat and drop them off — legends! In the restaurant, we garnish these with micro-fennel. It's difficult to get, but if you want to go the extra mile, it's worth it.

Shuck the oysters, removing the meat into a sieve to drain any excess liquid (this will keep the oysters crispy and you can use the juice to make a Dirty Oyster Martini, see p. 172). Keep the cupped (not the flat) part of the shells, and arrange on a serving plate. You can use a stack of rock salt to steady the shells, or some seaweed if you can get hold of it.

Ideally, use a deep-fat fryer and heat the oil to 190°C/375°F. You can also use a deep, heavy-bottomed pan or wok, but take care with the hot oil.

Coat the oysters in the flour and fry in the hot oil for 2 minutes. Remove with a slotted spoon onto paper towels to drain.

To serve, add a teaspoon of garlic crème fraîche into each oyster shell. Place a deep-fried oyster on top of the crème fraîche and garnish with a sliver of spring onion and some micro-fennel, if you have it.

Serves 4

12 meaty and plump oysters (Porthilly, if possible)
vegetable oil, for frying
plain (all-purpose) flour, for dredging
1 quantity Garlic Crème Fraîche (see p. 27)
2 spring onions (scallions), sliced into thin slivers and kept in ice-cold water until ready to serve, for garnish
micro-fennel, for garnish (optional)

Deep-fried Porthilly Oysters with Garlic Crème Fraîche

Pairs well with
Sparkling Albariño
Prosecco

Smoked Haddock Scotch Egg with Romesco Sauce

Pairs well with
Pinot Noir
Lager

Everyone loves Scotch eggs; well, proper ones, not those supermarket mass-produced dry things, but the ones you cut open to reveal a runny yolk. One of only a few labour-intensive recipes in this book, it's all in the preparation. These can be made a day or so in advance, and are bound to impress.

For the romesco sauce, preheat the oven to 160°C fan/180°C/350°F/gas mark 4.

Spread the almonds and bread over a baking tray (cookie sheet) and place in the oven for 8–10 minutes, to toast the nuts and dry the bread out. Remove and allow to cool.

Add the almonds, bread and all the remaining sauce ingredients to a food processor and blend until smooth. You may need to add some extra olive oil. Set aside.

For the Scotch eggs, bring a saucepan of water to the boil and carefully drop the 4 whole eggs in. Boil for 6 minutes, then remove immediately into iced water to stop the cooking process.

Add the parsley and smoked haddock to a food processor and

blend to a paste. Remove and divide the mixture into 4 portions. Roll each into a ball, then flatten out on a non-stick surface (greaseproof (wax) paper works well). If your mixture breaks up, add a little olive oil to moisten.

Remove all the shells from the soft-boiled eggs and place each onto a circle of haddock paste. Carefully wrap the eggs in the paste, completely covering them.

Prepare 3 dishes: one filled with flour, one with beaten egg and the last with breadcrumbs. First, roll the wrapped eggs in flour to cover, then roll in the beaten egg, then finally roll in the breadcrumbs, ensuring they are all completely covered.

In a deep saucepan, add a depth of 8 cm/3¼ inches of vegetable oil and heat over a high heat to 190°C/375°F. Add the breadcrumbed eggs and cook for around 5 minutes. You will need to gently turn the eggs to ensure they are nice and golden all over. You could also do this in a deep-fat fryer, if you have one.

Spoon the romesco sauce onto a serving dish and place the Scotch eggs on top. Alternatively, serve the sauce on the side.

Serves 4

For the romesco sauce:
80 g/2¾ oz/1 cup flaked (slivered) almonds
1 slice of stale white bread
100 g/3½ oz roasted red (bell) peppers
2 garlic cloves
1 tsp paprika
1 tbsp sherry vinegar
3 tbsp extra-virgin olive oil

For the Scotch eggs:
5 large free-range hen's eggs: 4 whole, 1 beaten
iced water, for cooling
a handful of parsley
400 g/14 oz undyed smoked haddock, skin removed
plain (all-purpose) flour, for dusting
100 g/3½ oz/scant 1 cup panko breadcrumbs
vegetable oil, for frying

Butterflied Sardines with Mango Salsa

Sardines are such an awesome species to cook — so quick, cheap, tasty and sustainable. Make sure they are as fresh as possible.

The mango salsa transports me straight back to sitting round the table with my parents and brothers. Mum used to make this when she served fajitas and, coupled with the oiliness of the sardines, it's bang on!

Place the sardines on a baking tray (oven pan), skin-side up, drizzle with olive oil and season.

Mix the mango, red onion, chilli, coriander and lime juice together with a drizzle of olive oil. Set aside.

Heat a drizzle of olive oil in a non-stick frying pan over a medium-high heat. Place the butterflied sardines into the pan, skin-side down. The meat of the fish will gradually change colour from a pinkish red to an opaque white. When it gets halfway up the side of the fish, flip the sardines over and cook for a further 30 seconds.

You can also cook these on a barbecue. Once it is up to temperature, oil the rungs of the barbecue with olive oil, then oil and season the skin of the fish and follow the same method as for cooking in a pan.

Transfer to a serving plate, spoon on a pile of the mango salsa and garnish with the coriander leaves.

Serves 4

12 sardines, butterflied (ask your fishmonger to do this)
olive oil, for drizzling
sea salt and freshly ground black pepper
½ mango, peeled, stoned and cubed
½ small red onion, finely diced
1 fresh red chilli, deseeded and finely sliced
a small bunch of coriander (cilantro), roughly chopped, with some leaves reserved for garnish
juice of 1 lime

Pairs well with
Vinho Verde

Crispy Falmouth Bay Shrimp with Sriracha Crème Fraîche

We don't have any large native prawns around the shores of the UK, so we have to import ours from further afield. Many people aren't even aware that we do have this fantastic species of British shrimp; although not very large, they are super-tasty. When deep-fried you can eat the whole thing, head and all. Any good fishmonger should be able to source these for you in season from late summer to early autumn, with some notice. Please give them a go!

This recipe needs to be cooked and served immediately, so have everything ready to go.

Ideally, use a deep-fat fryer (if you haven't got one, use a heavy-based saucepan) and heat the vegetable oil to 180–190°C/350–375°F.

Add the shrimp to a mixing bowl, sprinkle with seasoned flour, and turn with your hands, ensuring the shrimp are completely coated. Transfer to a sieve to remove any excess flour.

Now, carefully pour the shrimp into the hot oil and fry for 2–3 minutes, until crispy. Remove with a slotted spoon and place on paper towels to absorb any excess oil.

Transfer to a serving dish and sprinkle with a little sea salt. Mix the crème fraîche and sriracha together and serve on the side as a dipping sauce, along with lemon wedges to squeeze over the shrimp.

Serves 4

vegetable oil, for frying
200 g/7 oz Falmouth Bay/Mylor shrimp (small shrimp)
plain (all-purpose) flour, seasoned with sea salt and freshly ground black pepper, for dredging
sea salt, to taste
4 tbsp crème fraîche
1 tbsp sriracha chilli sauce
lemon wedges, to serve

Pairs well with
A Cornish white wine, such as Knightor Trevannion

Corn on the Cob and Brown Shrimp with Bone Marrow and Anchovy Butter

It's such a great time of year when corn on the cob is in season. It's available all year round these days, but for the best quality and flavour, buy British-grown corn in the summer months. Using bone marrow sounds complicated, but it's super simple and adds an extra richness to the dish.

Preheat the oven to 160°C fan/ 180°C/350°F/gas mark 4.

Place the marrow bone on a roasting pan and cook in the hot oven for about 10 minutes. Remove from the oven and set aside to cool slightly.

Once the marrow bone is cool enough to handle, scoop out the marrow and add to a food processor, along with the butter, anchovy fillets and a generous drizzle of the oil reserved from the anchovy can. Blend the mixture until smooth. Using a plastic spatula, scrape the butter from the food processor into a bowl, cover with cling film (plastic wrap) and store in the fridge until needed.

Bring a large pan of lightly salted water to a gentle simmer. Make sure the pan is large enough to fit the corn in. Add the corn cobs and simmer for about 5 minutes.

Meanwhile, add half of the bone marrow and anchovy butter to a small pan set over a medium heat. When the butter has melted, add the shrimp and fry for 2–3 minutes.

Drain the corn and transfer to a serving dish. Pour over the shrimp and butter mixture, and sprinkle with a little smoked paprika. Serve with the remaining butter in a dish on the side, for anyone who likes extra buttery corn!

Serves 4

For the bone marrow and anchovy butter:
1 beef marrow bone, split lengthways
100 g/3½ oz/7 tbsp unsalted butter, at room temperature
2 salted anchovy fillets (reserve some of the oil from the can)

4 corn cobs, husk removed if necessary
90 g/3¼ oz peeled brown shrimp (miniature shrimp)
pinch of smoked paprika

Pairs well with
Vinho Verde

Before I found my love of cooking, I spent 6 months living in China. There, I experienced some incredible food, unlike any other food I'd ever eaten. The use of Szechuan pepper, with its slightly numbing effect, is one of the distinctive flavours that etched itself upon my mind. Making a batch of this salt flavouring gives you a great little store cupboard secret weapon. It can be used on chips, deep-fried potatoes, crispy squid or when searing tuna.

To make the Szechuan salt, heat a frying pan (skillet) over a medium heat and toast the Szechuan pepper and black peppercorns for about 1 minute, to release their aroma. Transfer to a spice grinder and blitz thoroughly. Alternatively, crush as finely as possible in a pestle and mortar. Finally, combine with the Chinese five spice in a non-reactive (glass, ceramic or stainless steel) bowl and set aside.

Add the table salt to the frying pan, increase the heat to high, and cook for about 2–3 minutes, stirring occasionally, until the salt turns slightly grey (although a colour change is not essential). Tip the hot salt into the bowl with the rest of the spices and mix, to fuse the flavours together. Set aside and allow to cool. Kept in an airtight container, this salt will keep for about 6 months.

Bring the frying pan back up to a medium heat and drizzle in a little olive oil. Place the prawns in the pan, drizzle a little more olive over them and generously sprinkle about 2–3 tbsp of the flavoured salt over the prawns. Cook for about 2 minutes, until the prawns have changed from their natural grey colour to pink on the underside, then flip them over, sprinkle with a little more Szechuan salt and cook for a further 2 minutes. Transfer to a serving platter and garnish with the limes.

Serves 4

For the Szechuan salt:
1 tbsp Szechuan pepper
2 tbsp black peppercorns
4 tbsp Chinese five spice
9 tbsp table salt

For the prawns:
olive oil, for frying
12 raw tiger prawns (jumbo shrimp)
2 limes, halved

Szechuan Prawns

Pairs well with
Lager
Merlot

Stuffed Mussels with Capers, Garlic and Parsley

I'm not going to lie, this is a slightly labour-intensive recipe, but it's one that can be done well in advance of mates coming over and the mussels are very quick to cook. It's such a great way to eat these tasty molluscs.

Heat a large saucepan over a medium heat. Add the mussels, with a splash of water, and cover with a tight-fitting lid. Steam for about 3–4 minutes, until the mussels pop open. Tip into a colander, discarding any that have remained closed, and allow to cool.

Put the bread into a food processor and blend into fine crumbs. Add the butter, garlic, lemon zest and juice, Parmesan, parsley, capers and the white wine vinegar to the crumbs and blend again, until everything is well mixed.

Break off one half of the mussel shell from each mollusc and cover the meat of the mussel with a teaspoon of the stuffing mixture. Place them all on a baking tray (oven pan). If not cooking immediately, cover with cling film (plastic wrap) and store in the fridge.

When ready to cook, heat the grill (broiler) to high and, placing the tray quite close to the grill, grill (broil) the stuffed mussels for 2–3 minutes, until the mixture has turned a golden brown and the butter has melted. Transfer to a serving platter and get stuck in!

Serves 4

36 large live mussels, beards removed
2 thick slices of day-old bread, torn into pieces
150 g/5½ oz/⅔ cup unsalted butter
6 garlic cloves, roughly chopped
zest of 1 lemon and the juice of ½
2 tbsp grated Parmesan cheese
a medium bunch of parsley, roughly chopped
a small handful of capers, roughly chopped
a splash of white wine vinegar

Pairs well with
A lemony white wine, such as Albariño

Pan-fried Whole Squid with Capers, Preserved Lemon and Tarragon

Keeping squid whole and frying in the pan gives you a great contrast in textures – the wings crisp up and the body should melt in your mouth. Combine this with the dressing in this recipe and you'll find it's a match made in heaven! When buying your squid, ask your fishmonger to keep the wings of the squid attached – these parts are usually thrown away, but when pan-fried have a great flavour.

Mix together all the dressing ingredients and set aside.

Lay the squid out, so the wings are flat on the board. Using a sharp knife, carefully score the top side of the squid, ensuring these scores don't go through to the underside of the hood. Pat the squid and tentacles dry with paper towels, drizzle with olive oil and sprinkle with a little salt.

Heat a frying pan (skillet), large enough to hold all the squid, until nice and hot. Add the tentacles first as these need a little more cooking than the body. After 1 minute, lay the body in the pan with the wings flat on the surface of the pan. Don't be tempted to move the squid, just let it cook for about 2 minutes, until it has built up some golden colour, then flip it over and repeat.

Place the squid on the serving plates and spoon over the dressing, having given it a good mix beforehand. Add another sprinkle of salt and serve, with lemon wedges on the side.

Serves 4

For the dressing:
1 tbsp capers, chopped
2 tbsp white balsamic or sherry vinegar
a small handful of tarragon, chopped
1 tbsp preserved lemon, chopped
2 tsp of liquor from the preserved lemon jar
3 tbsp extra-virgin olive oil
1 garlic clove, finely chopped
1 banana shallot, finely chopped

4 squid, cleaned and kept whole, including tentacles, around 800 g/1lb 12 oz in total
extra-virgin olive oil, for drizzling
sea salt, to taste
wedges of lemon, to serve

Pairs well with
A dry, pale and crisp rosé from Provence or the Loire Valley

Crab-stuffed Courgette Flowers

This recipe combines the produce from two of our best suppliers. Ross has a kitchen garden just outside Padstow, and grows incredible veg; Johnny fishes crab and lobster for us just out of the Camel Estuary. Both have been a huge support to us, as well as becoming good friends. Understated in appearance (the dish, not Ross and Johnny!), it takes just one bite and you'll not want to share!

In a mixing bowl, add the crab meat, crème fraîche, spring onions and lemon juice and mix together, seasoning to taste.

Gently open the petals of the courgette flowers and spoon equal amounts of the crab mixture in between the petals, packing the mix in tightly and leaving enough space to be able to twist the petals back together. Ensure there are no gaps for the crab to escape out of.

Ideally, use a deep-fat fryer (if you haven't got one, use a heavy-based saucepan) and heat the vegetable oil to 190°C/375°F. Test the temperature by putting a cube of bread in the oil – if it immediately starts to crisp up, you're ready to go.

Meanwhile, make the tempura batter. Mix the plain flour, cornflour and baking powder together and slowly add the chilled sparkling water, whisking as you go. You're aiming for the consistency of double cream (you may need to add more or less than the quantity given – be guided by the consistency).

Gently coat the stuffed flowers in the batter and, using a spoon for support, slowly lower a flower into the hot oil, flower-end first. After a few seconds, let it submerge fully and fry for about 1½ minutes, until golden. Repeat for all 4 flowers.

Remove with a slotted spoon and place on paper towels to absorb any excess oil. Season with sea salt and they're ready to serve.

Serves 4

150 g/5½ oz white crab meat (unpasteurized)
1 tbsp crème fraîche
2 spring onions (scallions), finely sliced
1 tsp lemon juice
sea salt and freshly ground black pepper
4 courgette (zucchini) flowers, with the baby courgettes still attached
1 litre/35 fl oz/4⅓ cups vegetable oil

For the tempura batter:
75 g/2½ oz/½ cup plain (all-purpose) flour
50 g/1¾ oz/½ cup cornflour (cornstarch)
1½ tsp baking powder
approx. 150 ml/5 fl oz/scant ⅔ cup chilled sparkling water (you may need more or less)

Pairs well with
Chablis
Chardonnay

Whole Crispy Red Mullet with Lemon and Olive Oil

I had this very dish on the Greek island of Corfu, sat at plastic tables and chairs, looking across the bay with Katie and my parents. We all agreed that this dish was something special, both in flavour and simplicity. Whenever I eat it, I'm transported back to that trip.

A deep-fat fryer is the best thing to use for this recipe. If you don't have one, heat vegetable oil in a frying pan (skillet) large enough to fit the 4 fish. The oil needs to be deep enough to submerge half of the fish when laid on their sides. Bring the oil up to 190°C/375°F. Test the temperature by putting a cube of bread in the oil – if it immediately starts to crisp up, you're ready to go.

Pat the fish dry with paper towels, then dust with flour, ensuring they are completely covered and any excess is shaken off. Carefully place the fish in the hot oil. Fry for 4 minutes on each side. Remove with a slotted spoon onto some paper towels to absorb any excess oil.

Transfer to a serving plate, sprinkle with sea salt and serve with lemon wedges and the dressing in a small bowl on the side.

Serves 4

vegetable oil, for frying
4 red mullet, around 300 g/
    10½ oz each, scaled and gutted
    (gurnard can be used as an
    alternative)
plain (all-purpose) flour, for dusting
sea salt, to taste
lemon wedges, to serve
approx. 3–4 tbsp Extra-virgin Olive
    Oil and Lemon Dressing (see p. 25)

Pairs well with
English, Greek or
Italian lager
A dry crisp rosé
wine

Palourde Clams with Manzanilla and Garlic

Combining good-quality sherry with the amazing liquor from steamed clams creates an awesome, tasty sauce to mop up with some fresh crusty bread. The aroma will send you wild!

In a medium saucepan, set over a medium heat, add a little olive oil and cook the garlic and shallots until softened. Add the clams to the pan, along with a generous glug of Manzanilla sherry. Cover the pan with a lid and cook for about 1–2 minutes, until the clams open. Discard any that have not opened. Transfer to a serving dish and garnish with the parsley. Serve with some crusty bread on the side.

Serves 4

extra-virgin olive oil, for cooking
4 garlic cloves, finely chopped
2 banana shallots, finely chopped
1.2 kg/2 lb 12 oz live palourde/surf clams
a generous glug of Manzanilla sherry
finely chopped parsley, to garnish

Pairs well with
A crisp white wine, such as Muscadet

Cooking this recipe gets the senses going — the moment it goes into the oven, it fills the room with an fantastic aroma and you can't wait to tuck in! Once you've mastered the curry marinade, you can use it for other fish, such as fillets of cod or salmon, or whole fish such as bass or bream.

For the curry marinade, heat the butter in a frying pan (skillet) over a medium heat, add the onion, garlic and chilli and fry for about 5 minutes, until softened.

Meanwhile, heat a separate dry frying pan over a medium heat and toast the cumin and coriander seeds and cloves for about 1 minute to release the aromas. Transfer to a spice grinder, or pestle and mortar, and grind to a powder.

Add the softened onion, garlic and chilli to a food processor, along with the ground spice mix, cinnamon, lemon juice and coriander. Add a pinch of sea salt and the olive oil, then blend for about 1 minute to a rough, textured paste.

Preheat the oven to 160°C fan/180°C/350°F/gas mark 4.

Roughly score the mackerel on both sides to allow the curry marinade to penetrate the flesh of the fish. Line a roasting pan with some greaseproof (wax) paper and drizzle over a little oil to stop the mackerel sticking. Place the mackerel onto the paper and spoon the curry marinade over the mackerel, evenly coating the fish on both sides.

Roast the fish in the hot oven for 10–12 minutes.

Transfer to a serving dish and scatter with the toasted peanuts and sliced spring onion.

Serves 4

For the coriander curry:
1 tbsp salted butter
1 small white onion, roughly chopped
2 garlic cloves, finely chopped
1 fresh bird's eye chilli, finely sliced
½ tsp cumin seeds
½ tsp coriander seeds
¼ tsp whole cloves
¼ tsp ground cinnamon
juice of ½ lemon
1 medium bunch of coriander (cilantro)
pinch of sea salt
3 tbsp extra virgin olive oil

4 mackerel, gutted and cleaned (ask your fishmonger to do this for you)
2 tbsp crushed toasted peanuts, to serve
3 spring onions (scallions), finely sliced, to serve

Mackerel and Coriander Curry with Toasted Peanuts

Pairs well with
Riesling
Grüner Veltliner
IPA

Large Plates

There's nothing better than having one big dish to put down in the middle of the table for everyone to tuck into. For the perfect group meal, mix up some of the small tapas plates to start off with, then finish off with a showstopper from this section.

ıked

Turbot with M

e Roasted Mo

with Vietnam

Q Carabineros

Salt-baked Sea Bream

Salt baking fish is the ultimate recipe for a bit of dining theatre, as well as a great way of keeping the fish moist. You can experiment with different dry herbs to mix in with the salt to adjust the flavour.

Preheat the oven to 180°C fan/ 200°C/400°F/gas mark 6.

In a large mixing bowl, mix the salt, eggs and fennel seeds together to form a paste.

On a roasting pan, spread out half the salt mixture across the tray. Lay the fish on top, then stuff the cavity of the fish with half of the lemon wedges and all of the rosemary and thyme. Cover the fish with the remaining salt, ensuring the whole fish is completely covered, and pat

it down smoothly. Cook in the hot oven for 15 minutes.

Remove from the oven and leave the fish to rest for 5–10 minutes – this will allow the fish to continue cooking gently. After this time, give the salt crust a good bash with a rolling pin or the back of a knife. Use a large spoon to remove the chunks of salt, then carefully peel off the skin of the fish, trying not to let the salt touch the flesh.

You may want to serve the fish straight from the roasting pan, as it looks quite dramatic. Alternatively you can carefully transfer to a serving dish. Be sure to remove the skin, as it will be covered in salt. Serve with the remaining lemon wedges and a green salad.

Serves 4

2 kg/4 lb 8 oz coarse sea salt
4 free-range eggs
4 tsp fennel seeds
1 x 1.25 kg/2 lb 12 oz or 2 x 600 g/ 1 lb 5 oz sea bream (wild if available), scales on but gutted
2 lemons, cut into wedges
1 large bunch of rosemary
1 large bunch of thyme

Pairs well with
Picpoul de Pinet

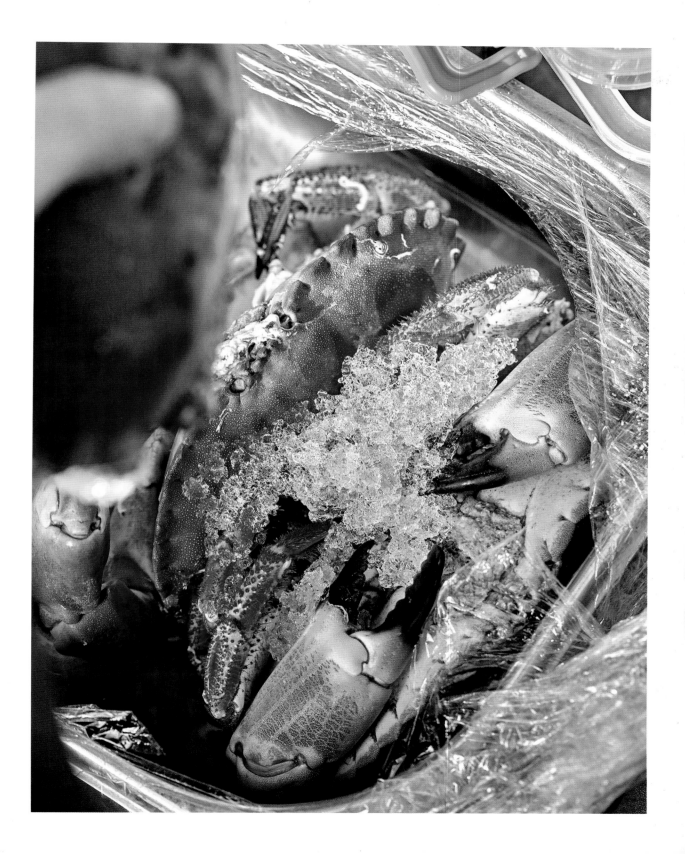

Roast Turbot with Mediterranean Vegetables and Salsa Verde

Pairs well with
A full-flavoured dry white wine, such as Meursault

Turbot is undeniably one of the tastiest fish around. It has to be cooked whole – the reason being that so much flavour comes from the bones. Roasting it on top of vegetables allows all the tasty juices from the fish to be soaked up by what's beneath. This is definitely a recipe for a special occasion, as turbot is one of the more expensive species, but it will impress.

Preheat the oven to 140°C fan/160°C/325°F/gas mark 3.
In a large roasting pan put the courgette, pepper, red onion, tomatoes and the unpeeled garlic cloves, drizzle with plenty of olive oil, and season well. Spread half the rosemary over the vegetables, add a splash of red wine vinegar and roast in the hot oven for 25 minutes.

Meanwhile, lay the fish out on a board and, using a sharp knife, score it down the middle, then make about 4 scores on each side, fanning outwards. Drizzle with olive oil, then tuck the peeled, sliced garlic into the scores along with the remaining rosemary. Season the fish and lay the lemon slices on top.

After they have roasted for 25 minutes, remove the vegetables from the oven. Turn the oven heat up to 160°C fan/180°C/350°F/gas mark 4. Place the fish on top of the vegetables and return to the oven to roast for a further 35 minutes.

Meanwhile, place all the ingredients for the salsa verde into a food processor and pulse until roughly blended.

Remove the fish from the oven, drizzle over the salsa verde and serve straight from the roasting dish. You could serve some nice crusty bread on the side to mop up the juices, or crushed spiced potatoes (see p. 150).

Serves 4

1 courgette (zucchini), roughly chopped
1 red (bell) pepper, roughly chopped
3 red onions, quartered
150 g/5½ oz cherry tomatoes, halved
8 garlic cloves, unpeeled, plus 4 garlic cloves, peeled and finely sliced
good-quality extra-virgin olive oil, for drizzling
sea salt and freshly ground black pepper
1 bunch of rosemary, leaves only
a splash of red wine vinegar
1.25–1.5 kg/2 lb 12 oz–3 lb 5 oz turbot, fins trimmed
1 lemon, sliced into rounds

For the salsa verde:
a small handful of basil
a small handful of mint
a small handful of flat-leaf parsley
1 garlic clove, peeled
4 anchovy fillets
2 tbsp capers
2 tsp red wine vinegar
6 tbsp good-quality olive oil

Covering a whole fish with this marinade is a sure-fire way of getting loads of flavour into any fish. It's not too overpowering and brings an awesome fresh, clean taste that you often don't get when eating out. This method works equally well on flat and round fish. Although not essential, when using flat fish, ask your fishmonger to remove the top layer of skin for you, to ensure the marinade flavours penetrate the meat. If the skin is still present, or if using round fish, just score the skin before marinating.

Preheat the oven to 160°C fan/ 180°C/350°F/gas mark 4. Line a roasting pan, large enough to hold the monkfish tail, with greaseproof (wax) paper.

Spread 2 tablespoons of the marinade across the greaseproof paper and place the monkfish tail on top. Make sure the meat of the butterflied monkfish is opened out and spread the remaining marinade over the fish. If you are using a different fish you may need more or less marinade – just make sure the fish is well covered.

Sprinkle with half the spring onion and roast in the hot oven for 20 minutes. When the fish is cooked, the meat will start to peel away from the backbone.

Transfer to a serving plate and pour the cooking juices over the fish. Garnish with the remaining spring onion, coriander, basil and peanuts (if using). Serve with lime wedges on the side.

Serves 4

8 tbsp Vietnamese Marinade (see p. 24)
1.5 kg/3 lb 5 oz monkfish tail on the bone, skinned and butterflied (ask your fishmonger to do this)
4 spring onions (scallions), finely sliced
4 sprigs of coriander (cilantro), leaves only
4 sprigs of holy basil (Thai basil), leaves only
a handful of toasted peanuts, chopped (optional)
1 lime, cut into wedges

Whole Roasted Monkfish Tail with Vietnamese Marinade

Pairs well with
Pinot Gris
Riesling

Johnny Murt's Crab Adobo

When we first started coming to Padstow to visit Johnny Murt, one of our suppliers who quickly became a great friend, he cooked up his signature dish of Chicken Adobo for us. Although renowned in the town, I wasn't so sure, looking at the amount of soy and cider vinegar going into the pot. However, when we came to eat it, I was amazed! Given that Johnny is a crab fisherman, I suggested we tried it with whole crab next time, and it worked so well. You can either use whole crab and pour the sauce over, or use crab claws, mixing it all in together. Make sure you serve a nice fresh salad with it (Johnny is a salad dodger!).

Heat a small saucepan over a medium heat, add a drizzle of olive oil, then add the garlic and ginger, cooking for about 2–3 minutes, until softened. Add the soy sauce, cider vinegar and bay leaves and leave to simmer for 15 minutes.

Meanwhile, preheat the oven to 160°C fan/180°C/350°F/gas mark 4.

Arrange the crab claws on a roasting pan and put them in the oven to warm through. This should only take 4–5 minutes.

Carefully transfer the crab claws into a serving dish, as the shells will be hot. Pour over the hot adobo dressing and garnish with the spring onions, the mooli or radish matchsticks and plenty of lime wedges.

Serves 4

olive oil, for drizzling
5 garlic cloves, finely sliced
4 tsp finely chopped fresh ginger
100 ml/3½ fl oz/7 tbsp soy sauce
100 ml/3½ fl oz/7 tbsp cider vinegar
2 bay leaves
8 large crab claws
4 spring onions (scallions), sliced
100 g/3½ oz mooli (daikon), peeled and sliced into matchsticks, or sliced radishes
lime wedges, to serve

Pairs well with
Pinot Gris
Riesling
Lager

BBQ Carabineros Prawns with Olive Oil and Sea Salt

Barbecuing these Mediterranean prawns is a brilliant way of adding a smoky element to these sweet-flavoured beauties. If you can't barbecue them, you can always grill or fry them. Sucking the prawn heads is mandatory! It's one of the easiest recipes ever, but the prawns have such a beautiful natural flavour, they really don't need much adding to them.

When your barbecue is up to temperature, drizzle the prawns with olive oil and season generously with sea salt. Lay the prawns on the barbecue, along with the lemon halves, and grill for 2 minutes, then flip the prawns over and grill for another 2 minutes. Check that the lemon halves are nicely charred.

Transfer to a serving plate and drizzle a little extra-virgin olive oil over them and season with salt and pepper. Garnish with the charred lemon.

Serves 4

12 raw carabineros prawns (jumbo shrimp)
olive oil, for grilling
2 lemons, halved
good-quality extra-virgin olive oil, for drizzling
sea salt and freshly ground black pepper

Pairs well with
Fino sherry
Cava

For the last few years, I've been lucky enough to be involved in a trip to Italy organized by our wine supplier, Tim. In a Tuscan hill-top town called Panzano, I had the ultimate meat experience... At Antica Macelleria Cecchini, where meat is life, the whole meal was amazing, but I just couldn't stop eating the whipped lardo! Added to whole roasted fish, it brings an amazing richness to the flavour.

For the truffled lardo, add the lardo, garlic, herbs and sea salt to a food processor and blend. Slowly pour in the truffle oil and keep blending until a light paste forms. You may have to remove the lid and push the lardo back into the bowl of the processor. If the lardo has not formed a light paste, add a little olive oil to loosen.

Preheat the oven to 160°C fan/180°C/350°F/gas mark 4. Line an oven tray, large enough to fit the fish and onions alongside each other, with greaseproof (wax) paper.

Score the skin on top of the brill, drizzle with olive oil and tuck the garlic slices into the scores. Drizzle olive oil over the red onions, splash with a little white wine vinegar and season well. Cover the fish with foil and roast in the hot oven for 30–40 minutes. Remove the foil 5 minutes before the end of this cooking time and spoon over the whipped lardo. Leave the fish uncovered for these final 5 minutes.

Transfer to a serving dish or serve straight from the roasting pan.

Serves 4

For the whipped truffled lardo:
150 g/5½ oz lardo (cured pork fat), skin removed and diced
1 garlic clove, peeled
1 tsp dried rosemary
1 tbsp dried thyme
a generous pinch of sea salt
1½ tbsp white truffle oil
olive oil (optional)

1.5 kg/3 lb 5 oz whole brill, fins trimmed (if you're feeling extravagant, turbot could also be used)
good-quality extra-virgin olive oil (Italian, if possible)
3 garlic cloves, finely sliced
4 small red onions, halved
a few splashes of white wine vinegar
sea salt and freshly ground black pepper

Whole Brill with Truffled Lardo and Red Onions

Pairs well with
Mâcon-Villages
A rich white Burgundy, such as Chablis

My first job when I moved to London was working front of house at a well-known seafood restaurant, previously owned by Mitch Tonks. The stand-out dish, and one that has stayed with me ever since, was the seafood stew. Sadly for me, Mitch was no longer involved in the restaurant when I worked there, so I never had the chance to learn it directly from him. This is my version of his recipe. I've cooked this many times for friends and family and it's such a sociable dish that gets everyone diving in.

In a food processor, combine all the parsley oil ingredients, blend well, and set aside.

For the stew, heat the olive oil in an extra-large lidded saucepan set over a low heat. Add the shallots, chilli, garlic and thyme, and sweat for about 2–3 minutes, until soft. Add the Pernod and cook for a further 3 minutes to cook off the alcohol. Add the fennel and cook for about 5 minutes, until soft. Add the tomatoes and saffron and cook for another 10 minutes.

Add the wine and fish stock. (If you want to add any steaks of fish, such as ray, bass, bream or gurnard, you could add them at this point.) Cook for 4 minutes, then add all the shellfish and cover with a lid to allow it to steam.

Once the mussels and clams have popped open, about 4–5 minutes, drizzle with the parsley oil and chopped basil and bring the pot to the table to serve. Some crusty bread is a must.

Serves 4

For the parsley oil:
100 ml/3½ fl oz/7 tbsp olive oil
1 garlic clove
3 sprigs of flat-leaf parsley

For the stew:
8 tbsp olive oil
2 shallots, finely chopped
1 fresh red chilli
4 garlic cloves, finely chopped
10 thyme sprigs, leaves only
1 fennel bulb, chopped
6 medium-sized tomatoes, quartered and roasted (roasting is optional)
2 pinches of saffron
a generous splash of Pernod
200 ml/7 fl oz/scant 1 cup white wine
150 ml/5 fl oz/scant ⅔ cup fish stock
600 g/1 lb 5 oz live mussels
350 g/12 oz live clams
4 langoustines
4 raw tiger prawns (jumbo shrimp), heads and shell on
a small handful of basil leaves, chopped, to garnish

Shellfish Stew with Parsley Oil

Pairs well with
Muscadet
Albariño

Whole Crab with Lime and Coriander Butter

This dish can be prepared in the oven, but it's also a great dish to do on the barbecue. We typically use this butter with crab and lobster, but you could also use it with crab claws, langoustines, or skinned flat fish such as Dover sole. The lime really cuts through the heaviness of the butter. It's a lovely social dish for everyone to tuck into.

Ask your fishmonger to remove the 'dead man's fingers' from the crabs and to split the bodies in half and crack the claws for you.

Preheat the oven to 160°C fan/ 180°C/350°F/gas mark 4, or get your barbecue up to temperature.

Place the crabs on a baking tray (oven pan) and put them into the hot oven. Cook for about 5 minutes, without turning. Alternatively, cook on the barbecue (not on the hottest part, or the shells will burn), turning halfway through, for 3–4 minutes on each side, until cooked through.

In the meantime put the butter, lime juice and salt and pepper into a small saucepan. Over a low heat, stir the butter constantly with a whisk, until the butter and lime juice have completely emulsified, leaving you with a smooth and silky sauce.

The crabs should be ready by the time the sauce is finished. Remove the crabs from the oven or barbecue (careful – they'll be hot!) and stack them up on a serving platter.

Throw most of the chopped coriander into the sauce (reserving a little for garnish), give it a good stir, then pour over the crabs. Scatter over the reserved coriander and spring onions, if using, and garnish with the lime halves.

Serves 4

2 cooked crabs, about 1 kg/ 2 lb 4 oz each
125 g/4½ oz/generous ½ cup unsalted butter, cubed
juice of 1 lime
pinch of sea salt and freshly ground black pepper
a small handful of chopped coriander (cilantro)
2 spring onions (scallions), finely sliced (optional)
1 lime, halved

Pairs well with
Pinot Gris
Gavi

Lobster with Smoked Chorizo Butter

Using chorizo is an amazing way of getting a slightly meaty, smoky flavour into the fish or shellfish you're cooking with. As with the previous recipe, it's great on crab and lobster, but you can also experiment with other species of fish. This lobster is perfect when served with the Tomato and Tarragon Salad (see p. 154) and/or our Crushed Spiced Potatoes (see p. 150). You can get smoked butter from delis or online.

Heat a saucepan or frying pan (skillet) over a medium heat and add a drizzle of olive oil. Add the chorizo and fry gently for 3–4 minutes, until it releases its oils.

Meanwhile, preheat the oven to 160°C fan/180°C/350°F/gas mark 4, or get your barbecue up to temperature.

Place the split lobster body and claws on a baking tray (oven pan) and put into the hot oven for 5 minutes. Alternatively cook on the barbecue (not on the hottest part, or the shells will burn), turning halfway through, for about 2 minutes on each side, basting with olive oil.

By now, the chorizo should have softened nicely, so add the smoked butter, lemon juice and seasoning, reduce the heat to low, and stir until the butter has melted.

Take the lobsters out of the oven/ off the barbecue (careful – they'll be hot!), place on a serving dish and pour the chorizo butter over the lobster. Sprinkle over the chopped chives and serve.

Serves 4

a drizzle of olive oil
100 g/3½ oz cooking chorizo, finely diced
2 cooked lobsters, split in half and claws cracked
125 g/4½ oz/generous ½ cup smoked butter
juice of ½ lemon
pinch of sea salt and freshly ground black pepper
a small bunch of chives, roughly chopped

Pairs well with
Sancerre
A white Burgundy such as Chablis

Monkfish and Chorizo Stew

Monkfish is one of those fish that can convert any non-fish eater. With dense flesh and no awkward little bones to pick through, it suits a pairing with meat very well. This is an awesome dish to chuck in the middle of the table and let everyone delve in.

Heat a large saucepan over a medium heat and add a good drizzle of olive oil. Add the chopped rosemary, fennel, red onions, garlic and chorizo and cook until softened, around 10 minutes.

Add the red wine and cook for a further 10 minutes. Add the tomato purée, chopped tomatoes, smoked paprika and the Marmite, check the seasoning, and mix well. Reduce the heat slightly and simmer gently for 45 minutes.

The chorizo stew can be made a couple of days before you need to use it, as long as it is kept in the fridge. If you've done this, heat it up in a saucepan before you use it in the dish, as the fish will not cook evenly if it's cold.

Preheat the oven to 160°C fan/180°C/350°F/gas mark 4.

In a large roasting pan spread out the hot chorizo stew. Lay the monkfish over the top of the stew, adding another drizzle of olive oil and seasoning with salt and pepper. Spike the reserved rosemary sprigs into the fish and cook in the hot oven for 25–30 minutes.

To finish, garnish with some fresh tarragon leaves and another little drizzle of olive oil. Serve with roasted potatoes or thick crusty bread.

Serves 4–6

olive oil, for drizzling
a handful of rosemary, finely chopped, plus 4 sprigs reserved whole
1 fennel bulb, finely sliced
2 red onions, finely diced
10 garlic cloves, finely chopped
200 g/7 oz good-quality cooking chorizo, finely diced
300 ml/10½ fl oz/generous 1¼ cups red wine
1 tbsp tomato purée (tomato paste)
3 x 400 g/14 oz cans of good-quality chopped tomatoes
2 tsp smoked paprika
2 tsp Marmite (yeast extract)
sea salt and freshly ground black pepper
1.5 kg/3 lb 5 oz monkfish tail on the bone, skinned and butterflied (ask your fishmonger to do this)
tarragon leaves, to garnish

Pairs well with
Red Rioja (slightly chilled)

Roasted Ray Wing with Olives, Chilli and Agridulce

Chef Mitch Tonks and our wine supplier Tim introduced me to agridulce vinegar, an amazing ingredient that turns a simple dish into something very special. Ray (skate) wings are one of my favourite fish – the meat slides off the top of the cartilage in long strands and has a wonderful texture and flavour.

Preheat the oven to 160°C fan/ 180°C/350°F/gas mark 4.

Line a roasting pan with a sheet of greaseproof (wax) paper. Drizzle with a little of the olive oil and place the ray wing(s) on top.

In a mixing bowl, combine the remaining ingredients and mix thoroughly. This can be done in advance and kept in the fridge. If you do this, make sure you remove the mixture from the fridge 30 minutes before using, to allow it to come back to room temperature.

Spoon the mixture evenly over the wing(s) and roast in the hot oven for 20 minutes. The cooking time will be a little less if you are using smaller wings, as they will be much thinner. Here's a great trick that our head chef Rob taught me, to check if the wing is cooked: using a pair of tongs, gently twist the piece of bone at the thickest point of the wing. If it disconnects easily from the flesh, it's time to remove the wing from the oven.

Once cooked, transfer carefully to a serving dish and pour the cooking liquor over the wing(s). Sprinkle over a little sea salt and serve.

Serves 4

3 tbsp extra-virgin olive oil
1.5 kg/3 lb 5 oz ray (skate) wing
   (can be 1 large or several smaller
   wings)
150 g/5½ oz mixed pitted olives,
   finely sliced
1 fresh red chilli, split in half
   lengthways and finely sliced
2 garlic cloves, finely chopped
2 tbsp agridulce vinegar
   (alternatively use a sweeter white
   wine vinegar)
sea salt, to taste

Pairs well with
Alsace Pinot
Gris

I lived in China for 6 months in my days after university. Struggling to get a job in design, I finally got a placement ... on the other side of the world! One weekend, a colleague took me fishing near her home town. We took the fish to a roadside restaurant and they made us a dish similar to this – well, as close as I could match it anyway!

Ideally you want to buy a grey mullet that has been caught at sea, rather than an estuary-caught mullet, which can have a stronger earthy taste – ask your fishmonger.

Preheat the oven to 160°C fan/180°C/350°F/gas mark 4. Line a roasting pan with greaseproof (wax) paper.

Score both sides of the mullet's skin, about 1 cm/½ in apart, to allow the flavour to enter the fish. Place the fish on the lined dish and rub the garlic into the scores.

Mix the soy sauce and mirin together and pour over the fish, then sprinkle the Chinese five spice over. Cover with the strips of carrot, spring onion and sliced chilli and add a final drizzle of olive oil all over the fish. Cook in the hot oven for 20 minutes.

To finish, sprinkle over a little more Chinese five spice, and garnish with chilli and spring onion.

Serves 4

1.5 kg/3 lb 5 oz grey mullet, scaled and gutted (ask your fishmonger to do this)
2 garlic cloves, finely sliced
2 tbsp dark soy sauce
2 tbsp mirin
1 tbsp Chinese five spice
2 carrots, peeled and sliced into thin strips
2 spring onions (scallions), sliced into thin strips
1 fresh red chilli, sliced
olive oil, for drizzling

For the garnish:
1 tsp Chinese five spice
1 fresh red chilli, finely sliced
2 spring onions (scallions), sliced into thin strips

Chinese-style Grey Mullet

Pairs well with
Gamay
Grüner Veltliner

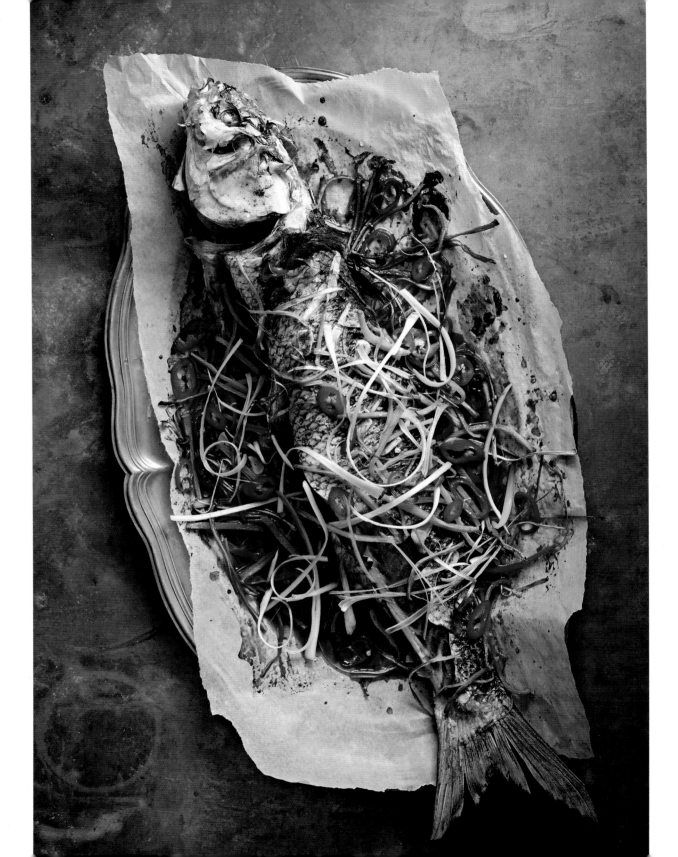

Sides

Side dishes should never be an after-thought and are just as important as the main event. They don't need to be labour-intensive or time-consuming; by combining just a few different ingredients you can create something that really complements the seafood you are serving. Here, we've put together our favourite sides from the last few years.

ed Spiced Pot

le Oil and Gra

ato and Tarro

ed Padrón Pep

Brown Shrimp with Asparagus, Pak Choi and Cashews

The brown shrimp in this recipe are optional, so the dish could be used as a veggie dish or as a side to accompany other fragrant dishes. We use Cornish asparagus from Ross and from St Enodoc, across the estuary from Padstow. It's only a short season, so when asparagus is available, we make the most it.

In a large saucepan, bring some lightly salted water to the boil, add the asparagus and cook for 30 seconds. Drain the asparagus, then immediately plunge into iced water. Pat dry with paper towels.

Combine the cooked asparagus, cucumber, pak choi, radishes, brown shrimp and nam jim dipping sauce in a mixing bowl and toss together thoroughly. Transfer to a serving plate, sprinkle the cashews and coriander over the top and garnish with the lime halves. Add sea salt, to taste.

If you prefer, you can switch the brown shrimp for prawns (shrimp), crab or even poached squid.

Serves 4

1 bunch of asparagus, woody stalks removed, sliced in half if large
iced water, for cooling
½ a cucumber, cut into ribbons with a peeler or a mandolin
1 pak choi (bok choy), cut into strips
6 radishes, finely sliced
100 g/3½ oz brown shrimp (miniature shrimp)
4 tbsp Nam Jim Dipping Sauce (see p. 26)
a small handful of toasted cashew nuts, crushed
a small handful of coriander (cilantro), chopped
2 limes, halved
sea salt

Pairs well with
Riesling

Crushed Spiced Potatoes

Rick doesn't like chips (controversial, we know!), but our customers were constantly asking for them. So, after playing around, this is what we came up with – a boiled potato/chip hybrid, that is more addictive than Pringles! We fry the potatoes for a crispy crunch – if you oven roast them, they won't be crunchy, although they will still be delicious.

This recipe makes a large quantity of spiced salt, but you can keep it in an airtight jar and use it whenever you like – it will keep for about 6 months.

In a large saucepan, add the potatoes, cover with cold water and bring to the boil. Parboil the potatoes for about 15 minutes, until soft, then drain and set aside.

Meanwhile, make the spiced salt. Heat a heavy-based frying pan (skillet) over a high heat and toast the cumin, coriander and fennel seeds for about 1 minute, until aromatic. Add them to a spice grinder and blitz, or crush in a pestle and mortar, then combine in a non-plastic mixing bowl with the other powdered spices.

Set the frying pan back over a high heat, add the salt and heat for about 2–3 minutes, until the salt turns a grey colour (although a colour change is not essential). Add to the mixing bowl and carefully combine the hot salt with the other spices (this helps the flavours of the spices to infuse the salt). Set aside to cool.

Gently crush the cooked potatoes on a board, under the palm of your hand, or use a potato masher to press them.

If you have a deep-fat fryer, heat the oil to 190°C/375°F. Fry the potatoes, in batches if necessary, for about 4–5 minutes, until golden. Drain on paper towels.

Otherwise, preheat the oven to 160°C fan/180°C/350°F/gas mark 4. Place the crushed potatoes in a large bowl and drizzle with olive oil, turning them to coat thoroughly. Spread the potatoes over a roasting pan and cook in the oven for 25–30 minutes, turning them every so often, until crispy and golden. Remove from the oven.

Sprinkle about 1–2 tsp of the spiced salt over the fried or roasted potatoes, giving them a good shake to ensure all the potatoes are well covered.

Place in a serving bowl and garnish with the chopped coriander and spring onion.

Serves 4

500 g/1 lb 2 oz new potatoes

For the spiced salt:
1 tsp cumin seeds
1 tsp coriander seeds
1 tsp fennel seeds
1 tsp smoked paprika (pimentón)
2 tsp ground turmeric
½ tsp ground cinnamon
100 g/3½ oz/⅓ cup table salt

vegetable oil, for deep-frying, or a drizzle of olive oil, for oven baking
a small handful of coriander (cilantro), chopped, to garnish
2 spring onions (scallions), thinly sliced, to garnish

Baby Gem Lettuce with Truffle Oil and Grana Padano

This side dish is super simple and mega tasty. The key is to use a good-quality truffle oil — it's really worth spending a little more to enhance this dish to its best.
If you can't find Grana Padano, you can substitute Parmesan.

Simply fan out the baby gem leaves on a serving dish or platter to enable the oil and cheese to reach all of the leaves. Drizzle the truffle oil all over, then sprinkle with the grated cheese. If you are really truffle-obsessive, you can drizzle a little more oil over the cheese. Grind black pepper over the top and serve immediately.

Serves 4

2 baby gem (Boston) lettuces, leaves
    separated, washed and dried
white truffle oil, for drizzling
4 tbsp shaved Grana Padano cheese
freshly ground black pepper

Tim, our good friend and wine supplier, introduced us to an amazing balsámico that is made close to Barcelona in Spain. This light and sweet vinegar takes this salad to another level. If you can't find the Ferret Guasch Balsámico that we use, any light sweet balsamic or sweetish sherry vinegar will work too. Make sure you buy tomatoes that are nice and ripe, preferably still on the vine.

In a mixing bowl, combine all the ingredients and mix thoroughly. Then simply transfer to a serving dish. Another option would be to arrange the salad leaves on a platter and spoon the tomato mixture over to serve.

Serves 4

350 g/12 oz vine-ripened tomatoes, roughly chopped
2 garlic cloves, finely chopped
3 sprigs of tarragon, leaves only, roughly chopped
3 tbsp good-quality extra-virgin olive oil
3 tbsp light, sweet balsamic vinegar
freshly ground black pepper
a generous pinch of sea salt
mixed salad leaves, to serve (optional)

Tomato and Tarragon Salad

Pattie's Soda Bread

I'm lucky enough to have two dads! Before we started Prawn on the Lawn, we'd come home to a house smelling of my stepdad Pattie's freshly baked soda bread. It's an awesome bread, as there's no proving needed, so from start to finish you can make it in just under an hour.

Preheat the oven to 160°C fan/ 180°C/350°F/gas mark 4. Grease a 900 g/2 lb loaf tin (pan) and dust lightly with a little flour.

Put the plain flour, wholemeal flour, bicarbonate of soda and salt into a large bowl and combine thoroughly. Add the buttermilk and mix with your hands until a dough forms. It's important to mix the dough as little as possible, to ensure a good rise.

Transfer the dough to the loaf tin and bake in the hot oven for 45 minutes, until golden. Remove from the oven, turn it out from the tin and leave to cool on a wire rack.

Serves 4

a little butter or oil for greasing the tin
115 g/4 oz/scant 1 cup plain (all-purpose) flour, plus a little for dusting the tin
340 g/11¾ oz/2¾ cups wholemeal (wholewheat) flour
1 tsp bicarbonate of soda (baking soda)
½ tsp table salt
450 ml/16 fl oz/2 cups buttermilk

Poppy Seed and Seaweed Flatbread

This is a fantastic accompaniment to the Whipped Cod's Roe (p. 54), Mackerel Pâté (p. 51) and Cured Salmon (p. 67) recipes. You can experiment with different dry toppings here, if you wish.

Preheat the oven to 160°C fan/180°C/350°F/gas mark 4.

Combine the plain and wholemeal flours with the milk in a mixing bowl and knead to a sticky dough. Using a rolling pin, roll out the dough on some greaseproof (wax) paper to a 2–3 mm/⅛ in thickness. Place the dough, on its paper, onto a baking sheet.

Brush the surface of the dough with the egg yolk, and sprinkle over the poppy seeds, seaweed and a pinch of salt.

Bake in the preheated oven for 6–7 minutes. Keep a close eye on it – you want a nice golden colour. Remove from the oven, leave to cool on the baking sheet, and snap into whatever size you want, to serve.

Serves 4

100 g/3½ oz/¾ cup plain (all-purpose) flour
100 g/3½ oz/¾ cup wholemeal (wholewheat) flour
3 tbsp milk
1 egg yolk, whisked
2 tsp poppy seeds
2 tsp dried dulse seaweed
a pinch of salt

You wouldn't expect to find Padrón peppers in Padstow, but Ross Geach at the Padstow Kitchen Garden does a great job at growing them! As soon as these small green peppers come into season, we get them straight on the menu and they go down a storm. Beware though – it's said that 1 in 10 are super hot!!

In a frying pan (skillet) large enough to hold the peppers, heat the olive oil over a medium heat, until smoking hot. Very carefully add the Padrón peppers. Fry, keeping them moving in the pan. As soon as you see the skins start to blister, remove the peppers from the pan with a slotted spoon on to some paper towels. Transfer to a serving dish and sprinkle generously with sea salt flakes, to serve.

Serves 4

3 tbsp extra-virgin olive oil
24 Padrón peppers
sea salt flakes

Fried Padrón Peppers

With this dish, it's all about the ingredients! Asparagus has a very short season (around 6 weeks in the UK), so you really need to make the most of it while you can. We are lucky to have two fantastic suppliers within a couple of miles of the restaurant, so we get it super fresh: St Enodoc Asparagus and Padstow Kitchen Garden. This recipe deviates slightly from the English classic, but the addition of soy sauce really is a game changer and doesn't overpower the flavour of the asparagus.

Remove the woody base of the asparagus by either snapping or cutting it off. When you bend the asparagus, it tends to snap where the woody base meets the tender part of the stem.

To make the soy butter combine the butter, lemon juice and soy sauce in a small saucepan set over a low heat. Stir continuously until melted, then set aside.

Heat a large non-stick frying pan (skillet) over a medium heat. Drizzle the asparagus with olive oil, season with salt, then place into the pan and cover with a lid. Steam for 2 minutes, then turn the asparagus over, replace the lid and cook for a further 2 minutes.

If you need to, reheat the butter slightly. Transfer the cooked asparagus to a serving dish and either pour the butter over the top or serve the butter in a ramekin or small bowl on the side, for dipping.

Serves 4

a large bunch of asparagus, approx. 400 g/14 oz
100 g/3½ oz/7 tbsp unsalted butter
juice of ½ small lemon
1½ tbsp soy sauce
olive oil, for drizzling
sea salt, to season

Charred Asparagus with Soy Butter

Desserts and Cocktails

Prawn on the Lawn is all about the seafood, so desserts were somewhat of an afterthought in the restaurant, only making it onto the menu after protests from customers to have something sweet to finish! We just keep things simple and only offer a couple, along with a plate of local cheeses and homemade chutney.

d Ca

on Posset and

ched Apricots

rnish Negron

hai-style Bloo

r M

Life wouldn't be the same without a dessert to finish your meal off. This one is dangerous!

Melt the butter in a saucepan set over a medium heat. Add the sugar, the double cream, the vanilla and salt, and mix thoroughly. Bring the mixture to the boil, then reduce to a simmer for about 5 minutes, until it thickens slightly.

Place 4 amaretti biscuits into 4 small pots or ramekins (about 240ml/ 8 fl oz/1 cup capacity), and crush lightly with the end of a rolling pin. Carefully pour the hot caramel mixture over the biscuits and transfer to the fridge until needed.

When ready to serve, spoon a little crème fraîche into each pot, top with fresh raspberries and garnish with a sprig of mint.

Serves 4 (or 6 slightly smaller servings)

For the caramel:
125 g/4½ oz/generous ½ cup unsalted butter
210 g/7½ oz/generous 1 cup soft light brown (light muscovado) sugar
150 ml/5 fl oz/scant ⅔ cup double (heavy) cream
¼ tsp vanilla bean paste
2 pinches of sea salt flakes

To serve:
16 amaretti biscuits (cookies) (make sure they're the hard ones, not soft)
350 ml/12 fl oz/1½ cups crème fraîche
200 g/7 oz fresh raspberries (about 12 in total)
4 sprigs of mint

Salted Caramel Pot

Pairs well with
Monbazillac
Sauternes

Lemon Posset and Coconut Shortbread

This is a really simple, fruity and light dessert, ideal for those who say they don't like desserts. You can make the posset and the shortbread a day in advance, so it's really easy to put together at the last minute.

First make the possets. Place a saucepan over a low heat, add the cream and sugar and stir gently, until the sugar has melted. Bring to a simmer and let it bubble away for 1 minute. Take the pan off the heat and stir in the lemon zest and juice. Pour an equal amount into 4 small Kilner jars or ramekins and leave to cool to room temperature. Once cooled, carefully cover with a lid or cling film (plastic wrap) and chill in the fridge for at least 3 hours and up to 24 hours.

Next make the shortbread. Preheat the oven to 160°C fan/180°C/350°F/gas mark 4.

Spread the coconut out on a small baking sheet and place in the hot oven for about 5–10 minutes, until lightly toasted. Make sure you keep a close eye on it, as it can burn fairly quickly. Set aside to cool.

Line 2 large baking sheets with greaseproof (wax) paper.

Cream the butter and sugar together in a large bowl, using an electric or hand whisk, until light and fluffy. Sift in the flour and cornflour and add the toasted coconut. Mix to form a dough, then roll out to about 1 cm/½ in thick on a lightly floured surface. Use a biscuit cutter to cut out small biscuits and keep re-rolling and cutting until you have used all the dough. Place the biscuits onto the lined baking sheets and sprinkle with some extra coconut. Place in the fridge for about 30 minutes, to stop the shortbread spreading when cooking.

Bake the biscuits in the hot oven for 20 minutes, or until lightly golden. Remove from the oven and leave to cool on the baking sheets for 10 minutes, then transfer to a wire rack to cool completely.

Serve the lemon possets straight from the fridge and place the biscuits on a large plate in the middle of the table so that everyone can help themselves. Store any leftover biscuits in an airtight container – they are great with a cup of tea or coffee.

Serves 4

For the possets:
400 ml/14 fl oz/1¾ cups double (heavy) cream
135 g/4¾ oz/⅔ cup golden caster (superfine) sugar
zest of 2 lemons, plus 50 ml/1½ fl oz/3 tbsp juice

For the coconut shortbread:
50 g/1¾ oz/generous ½ cup desiccated (dried grated) coconut, plus extra for sprinkling
225 g/8 oz/1 cup butter, at room temperature (you could also use a mixture of butter and coconut butter)
110 g/3¾ oz/½ cup caster (superfine) sugar
200 g/7 oz/1½ cups plain (all purpose) flour
100 g/3½ oz/1 cup cornflour (cornstarch)
a pinch of salt

Pairs well with
Limoncello

This is a great dessert recipe with a little twist: Sweet apricots combine with the more savoury elements of the ricotta and walnuts.

Place the apricot halves in a saucepan with the sugar, vanilla pod seeds and the dessert wine. Ensure the apricots are covered by the liquid; if not, top up with a little water. Place over a low-medium heat and bring the liquor up to a gentle simmer. Poach until the apricots are soft; this should take around 20 minutes. Remove the fruit from the liquor and set aside.

Continue to simmer the liquor to reduce it to the consistency of double (heavy) cream; this should take around 10 minutes.

To serve, arrange the apricot halves in serving bowls. Place spoonfuls of the ricotta around the apricots, then drizzle with the cooking liquor and honey. Scatter the crushed walnuts over the top and garnish with the lemon verbena leaves.

Serves 4

400 g/14 oz fresh apricots, halved
   and stoned (pitted)
2 tbsp caster (superfine) sugar
seeds from 1 vanilla pod (or ¼ tsp
   vanilla bean paste)
200 ml/7 fl oz/scant 1 cup
   Monbazillac or other sweet
   dessert wine
200 g/7 oz soft ricotta
1 tbsp truffle honey
3 tbsp crushed walnuts
10 lemon verbena leaves (or the
   leaves from 4 lemon thyme sprigs)

Pairs well with
Monbazillac

Poached Apricots with Ricotta, Truffle Honey and Walnuts

Affogato

This is the simplest of dessert recipes, but is perfect when you want something sweet but not too heavy. You can make it boozy or not boozy – tailor it to the mood! If you're feeling indecisive, then you can put a shot of coffee and a shot of alcohol in.

Take 4 teacups or small bowls and put a scoop of ice cream in each. Serve with the hot coffee/coffee liqueur/sherry in little jugs on the side. Pour the shot over the ice cream and eat! Yes, it is that simple. We like to dunk an amaretti biscuit in the melted ice cream as an extra treat.

Serves 4

4 scoops of good-quality ice cream (vanilla is the classic, but salted caramel also works well)
4 shots of either freshly made espresso coffee or coffee liqueur (we use Bepi Tosolini, but you can also use Tia Maria or Baileys Irish Cream) or Pedro Ximénez (or similar sweet sherry)
4 amaretti biscuits (cookies)

Banoffee Pot

As a kid, Rick went through a stage of making a banoffee pie every Sunday and then having it for breakfast every morning throughout the week. Needless to say, he was a little bigger then than he is now! It's always tricky cutting a banoffee pie into neat equal pieces, so we decided to put it into individual pots to make it easier to serve in the restaurant.

First make the caramel. Put the butter, sugar and double cream into a saucepan and melt over a medium heat. Bring to the boil and boil for 5 minutes, until the mixture has thickened slightly. Stir in the vanilla bean paste, then remove from the heat and set aside to cool.

Just before you are ready to serve, crumble 1–1½ digestive biscuits into the bottom of 4 small Kilner jars or short glasses. Spoon the cooled caramel over the crumbled biscuits. Slice the bananas and lay several pieces over the caramel layer. Whip the cream to stiff peaks and fold in the Greek yogurt, then spoon this mixture over the top of the bananas. Top with the chocolate curls and serve immediately, so the bananas don't brown.

Serves 4

100 g/3½ oz/7 tbsp unsalted butter
170 g/6 oz/generous ¾ cup soft light brown (light muscovado) sugar
125 ml/4 fl oz/½ cup double (heavy) cream
¼ tsp vanilla bean paste
4–6 digestive biscuits (graham crackers)
2 medium bananas
4 tbsp whipping cream
110 g/3¾ oz/½ cup Greek yogurt
dark chocolate curls or shavings, to decorate

Cornish Negroni

Put plenty of ice cubes into a cut-glass tumbler. Add your liquid ingredients, stir a few times and then finish off with orange zest or a slice of orange. Add more ice cubes, if needed.

50 ml/1½ fl oz/3 tbsp Tarquin's Cornish Gin (or ordinary gin)
25 ml/¾ fl oz/1½ tbsp Campari
25 ml/¾ fl oz/1½ tbsp Cocchi Vermouth di Torino (or ordinary red vermouth)
strip of orange zest or a slice of orange

Classic Margarita

Add all the ingredients, except for the salt, to a cocktail shaker filled with a handful of ice and shake well. Strain into a glass over ice cubes and then serve with a slice of lime. Salting the rim of the glass first is optional – depends on what your guest wants!

50 ml/1½ fl oz/3 tbsp tequila
25 ml/¾ fl oz/1½ tbsp Cointreau
a squeeze of agave nectar (syrup)
the juice of 1 lime, plus 1 lime cut into slices for garnish
salt, for the rim of the glass (optional)

Thai-style Bloody Mary

Put everything into a glass over ice, topping up with enough tomato juice to fill, and stir. Add a wedge of lime, a celery stick and a stripy straw.

Add an oyster on top, if required. Take the Tabasco sauce to the table, to serve, and guests can add more, to taste.

25 ml/¾ fl oz/1½ tbsp vodka
2 tsp lime juice
a few shakes of Worcestershire
    sauce, or to taste
finely diced fresh red chilli, to taste
½ tsp chopped coriander (cilantro)
tomato juice, to top up

To garnish:
a wedge of lime
a stick of celery
1 fresh oyster (optional)
a few drops of Tabasco sauce

Pinkster Gin and Tonic

Fill a glass with ice. Add the gin and raspberries, then 'clap' the mint leaves between your hands to release their aroma and add on top. Serve the tonic on the side for guests to serve themselves.

25 ml/¾ fl oz/1½ tbsp or 50 ml/
    1½ fl oz/3 tbsp Pinkster Gin
2 raspberries
5–6 mint leaves
tonic water, to taste

Kir Royale

Pour a small amount of cassis into a coupe glass, then fill to the top with prosecco or Cornish sparkling wine. Add a raspberry or an edible flower, to garnish.

a splash of crème de cassis
prosecco or Cornish sparkling wine
1 raspberry or edible flower

Dirty Oyster Martini

Add the vodka, dry martini and oyster juice to a cocktail shaker with a handful of ice and shake well. Strain into a martini glass, and serve with the oyster and a lemon wedge on the side.

75 ml/2¼ fl oz/4½ tbsp vodka
15 ml/½ fl oz/1 tbsp dry martini
the juice from 1 oyster, keeping the
    oyster itself for garnish
lemon wedge, for garnish

Index

Suppliers

Scotland

Armstrong's of Stockbridge
80 Raeburn Place
Edinburgh EH4 1HH
www.armstrongsofstockbridge.co.uk

The Fish People
350 Scotland Street
Glasgow G5 8QF
www.thefishpeopleshop.co.uk

Welch Fishmongers
23 Pier Place
Newhaven
Edinburgh
EH6 4LP,
www.welchfishmongers.com

North East England

F.R. Fowler & Son
1 Gerard Avenue
Burnholme
York YO31 0QT
www.fowlersofyork.co.uk

Latimer's Seafood Ltd
Shell Hill
Bents Road
Whitburn
Sunderland SR6 7NT
www.latimers.com

Ramus Seafoods
Ocean House
Kings Road
Harrogate
North Yorkshire HG1 5HY
www.ramus.co.uk

Taylor Foods
19a Elm Road
West Chirton North Industrial Estate
North Shields NE29 8SE
www.taylorfoods.co.uk

North West England

Out of the Blue Fishmongers
484 Wilbraham Road
Chorlton-cum-Hardy
Manchester M21 9AS
www.outofthebluefish.co.uk

Wales

E. Ashton Fishmongers Ltd
Central Market
Cardiff CF10 1AU
www.ashtonfishmongers.co.uk

The Fabulous Fish Company
Newhall Farm Shop
Pwllmeyric
Chepstow
NP16 6LF
www.fabulousfish.co.uk

South West England

Field & Flower – online only
www.fieldandflower.co.uk

Fish for Thought
The Cornish Fish Store
Unit 1, Callywith Gate Business Park
Launceston Road
Bodmin
Cornwall PL31 2RQ
www.fishforthought.co.uk

Prawn on the Lawn
11 Duke Street
Padstow
Cornwall PL28 8AB
www.prawnonthelawn.com

The Cornish Fishmonger
Wing of St Mawes Ltd
Unit 4
Warren Road
Indian Queens Industrial Estate
Indian Queens TR9 6TL
www. thecornishfishmonger.co.uk

Rockfish Retail Fish Market,
Brixham Fish Market
Brixham,
TQ58A
www.therockfish.co.uk

South East England

A passion for Seafood
www.apassionforseafood.com

Any Fish Ltd
2 Merlin Mews
Houchin Street
Bishops Watham
 Southampton
 SO321AR
 www.anyfish.co.uk

The Fish Society
Fish Palace
Coopers Place
Godalming GU8 5TG
www.thefishsociety.co.uk

Veasey & Sons Fishmongers
17 Hartfield Road
Forest Row
East Sussex RH18 5DN
www.veaseyandsons.co.uk

London

Elias Fish
8 Bittacy Hill
Mill Hill
London NW7 1LB
www.eliasfish.co.uk

Faircatch – box scheme for South-
 West Londoners
80 Geraldine Road
London SW18 2NL
www.faircatch.co.uk

Fin and Flounder
71 Broadway Market
London Fields
Hackney
London E8 4PH
www.finandflounder.co.uk

Moxon's
Locations in Clapham South, South
 Kensington, East Dulwich and
 Islington
www.moxonsfreshfish.com

Prawn on the Lawn
292–294 St.Pauls Road
London N1 2LH
www.prawnonthelawn.com

The Chelsea Fishmonger
10 Cale Street
Chelsea
London SW3 3QU
www.thechelseafishmonger.co.uk

- ...OLE, A LA PLANCHA...
- WINKLES, GARLIC BUTTER £7
- SQUID, N'DUJA, TARRAGON
- RAY WING, WILD GARLIC, FEN...
- BURRATA, CRISPY ARTICHOKE
- SZECHUAN PRAWNS £8.5

- BABY GEM, TRUFFLE OIL, 6/
- TOMATO + TARRAGON SALA...
- HOME MADE SODA BREAD

Acknowledgements

There are so many people to thank for getting us to the point of publishing our first book. We'd love to dedicate this book to everyone who has helped along the way. If we've forgotten anyone, we're sorry, but we love you and are so grateful for everything!

Firstly, thanks to our parents, grandparents, my bros, Will and Harry, and cousin Will, who have been amazing in all areas of POTL and our lives in general – we couldn't have done any of this without you.

To all our friends, those who've helped both in the restaurant and out, who've put up with us missing important occasions, and listened to us talk relentlessly about the restaurant and still chosen to hang out with us despite everything!

To our awesome staff, past and present, who work day in, day out, to carry out our vision for what the POTL experience should be. We're so grateful for all of your hard work, and we love working with each and every one of you. Daniela and Patrizia, thank you so much for your amazing work keeping the Islington restaurant running as if we were there ourselves – you guys are awesome. And Cam, Eddie, Jonny and Harriet in Padstow we wouldn't be where we are today without you all..

To Jamie, for cycling past the butchers and texting me a pic of the 'to let' sign, which became the home of the first POTL, and to Reg, our landlord in Islington, who gave us the opportunity to start our dream, we can't thank you enough. Chippy (Elsewhere Studio) thank you so much for our branding and design and being very patient with us over the years and Harry (Clear Honest Design and my bro) for all his magic with the website.

To our suppliers, who have provide us with the best ingredients, we hope we do your products justice. Special mention to the Murts Shellfish, Wild Harbour, Tim and Luke at Rock Shellfish, George Cleave a.k.a Mr Deluxe, Ross Geach (Padstow

Kitchen Garden), Liz and Engin (Trevibban Mill), Des and Caron (Padstow Brewing Co.), Damian Sais (Elias Fish). Thank you to Ron and Sonia (Trerethern Farm) for letting us set-up on your farm for the summer, you helped save our business!

To Mitch Tonks, for the foreword to the book, for being our inspiration and someone for whom we have a huge amount of respect – thank you so much for all your support.

To Tim and Alison (Sommelier's Choice), thank you for providing us with wicked wines, more support than we could imagine and all the awesome wine recommendations throughout the book.

To Paul and Emma Ainsworth and the team, for being so welcoming when we first came to Padstow. I had an incredible experience working in your kitchen. Thank you.

To all the guys at Pavilion and Emily Preece-Morrison, who believed in what we do, and gave us this opportunity to put together our favourite recipes to share with everyone, thanks for coping with us missing nearly every single deadline!

Finally, to Steven Joyce, Oliver Rowe, Lizzie Kamenetzky and Linda Berlin for the wicked pics throughout the book. We had so much fun with you and all the team – negronis on the beach at 8am should be mandatory!

Rick and Katie